Historical Variability in Heritable General Intelligence

Its Evolutionary Origins and Socio-Cultural Consequences

Historical Variability in Heritable General Intelligence

Its Evolutionary Origins and Socio-Cultural Consequences

Michael A. Woodley
and
Aurelio José Figueredo

The University of Buckingham Press

First published in Great Britain in 2013 by

The University of Buckingham Press
Yeomanry House
Hunter Street
Buckingham MK18 1EG

A CIP catalogue record for this book is available at the British
Library

ISBN 9781908684264

MICHAEL WOODLEY is an individual differences researcher and behavior geneticist who holds a post-doctoral fellowship with Umeå University and an honorary post-doctoral fellowship with the Center Leo Apostel for Interdisciplinary Studies, Vrije Universiteit Brussel. He lives in London, UK.

AURELIO JOSÉ FIGUEREDO is an individual differences researcher and evolutionary psychologist, who has dual professorships with the University of Arizona in the departments of psychology and family studies and human development. He is also director of the universities' ethology and evolutionary psychology graduate program. He lives in Tucson, Arizona.

CONTENTS

SUMMARY

Reconstructed changes in the mean levels of the heritable component of general intelligence (shortened to $g.h$) trend in the same direction as the Western historical social status-fertility relation (i.e. positively up until the early to mid 1800's and then negatively after) and are strong predictors of changes in the per capita rates of innovation in science and technology (Woodley, 2012a). A prediction from 'smart fraction' theory is that this effect should be strongly mediated by changes in the numbers of eminent individuals, i.e. those who actually generate innovation. This is tested using an index of fractionally scaled Western $g.h$ means, along with data on per capita innovation rates (Huebner, 2005), per capita eminent individuals (Murray, 2003) and also a measure of the fractionally scaled change in the level of environmentally sensitive, narrow cognitive abilities (shortened to $s.e$), which captures the Flynn effect. In testing the theory we employed a combination of growth curve and hierarchically nested path analysis in which serially autoregressive effects are considered. Consistent with smart fraction theory, the best fitting model reveals that eminent individuals mediate the impact of changing $g.h$ on innovation rates (where $\beta g.h \rightarrow$ Eminent individuals=0.915, and βEminent individuals$\rightarrow$Innovation rates=0.427), in addition to $g.h$ exhibiting a direct effect on innovation rates ($\beta g.h \rightarrow$ Innovation rates=0.495). Rising $s.e$ by contrast has had no effect on the increase and subsequent decrease in per capita rates of eminent individuals and significant innovations. This is consistent with the finding that dysgenic fertility is a Jensen effect in that it is significantly more pronounced on more g-loaded subtests (Woodley & Meisenberg, in press), unlike the Flynn effect, which is significantly more pronounced on less g-loaded subtests (te Nijenhuis & van der Flier, in press), indicating that dysgenesis is principally associated with a decline in *g.h* and therefore the capacity for significant innovation, whereas the Flynn effect is simultaneously associated with secular gains predominantly in the *s.e* component. This presents a solution to Cattell's paradox, or the contradiction presented by the simultaneous observations of negative correlations between IQ and fertility throughout the 20th century and the Flynn effect. These findings are evaluated in the context of a multilevel selection model, which proposes that innovation represents a 'eugenic' group-selected altruistic behavior due to the low individual benefits documented for the innovators and the huge collective benefits documented for the group (Cattell, 1972; Clark, 2007; Figueredo, 2012; Hamilton, 2000). One prediction of this multilevel selection model connects individual-level selection for increasing within-population mean *g.h* with group-level expansion facilitated via the application of innovations to the process of conquest and colonization, which in turn offsets the individual-level selection *against* those rare individuals possessing ultra-high IQs and other traits predisposing towards the generation of significant innovation. Another prediction connects the process of rising within group *g.h* with periods of heightened intergroup conflict facilitated by sources of ecological harshness. Conversely, reduced ecological harshness coupled with reduced inter-group conflict relaxes both individual and group-level selection for rising *g.h*, leading instead to individual-level selection for declining *g.h*. The dynamics of these evolutionary processes during the last half-millennium are presented as supportive of this model.

1. INTRODUCTION

One of the most controversial findings in intelligence research is the existence of negative correlations between IQ and fertility. As long ago as the 19th century, it was feared that civilization had gone too far in alleviating the Darwinian pressures keeping those with 'valuable' traits, such as high-intelligence, good health and character in high-fertility, whilst supressing the same amongst those endowed with low levels of these traits (e.g. Darwin, 1871; Galton, 1869). This fear was manifest despite the absence of either sophisticated psychometric and anthropometric measures of these 'valuable traits' or compelling data indicating that their levels had actually changed. Early in the 20th century, the term 'dysgenics' was coined to describe this phenomenon (Saleeby, 1913, 1914; Starr-Jordan, 1915).

Dysgenics was frequently defined in normative terms, with negative changes in the level of traits such as IQ being described as 'socially undesirable'. The term itself literally means 'bad-breeding'. The no less normatively loaded antonym of dysgenics, 'eugenics' (meaning good-breeding) was coined even earlier (Galton, 1883). The idea was to use eugenic measures to counter the socially undesirable consequences of dysgenic fertility.

When attempting a scientific discussion of issues such as dysgenics and eugenics, we believe that it is very important to define the relevant terms with greater precision than is typically done, as this field of research is fraught with many misunderstandings. In doing so, however, we will of necessity be making certain interpretations regarding the way that some of the more politically-charged terms have been used historically, and we will be providing evidence for those interpretations.

The first 'term of art' that needs clarification is *fitness*, as opposed to *selection*. *Selection* is a process and *fitness* is the outcome of that process. Fitness is defined in population biology as gene-copying success under a particular regime of selection, although it is sometimes used colloquially as the heritable phenotypic traits that confer such an advantage. As a consequence of the many misunderstandings of the word "fitness", we limit our use of that term in this monograph exclusively to instances where fitness is explicitly invoked as an outcome, and retain "selection" for where we describe the differentials in survival and reproduction produced by different selective processes, or *selective pressures*, such as within-group and between-group competition.

The total fitness (*f*), or gene-copying success, achieved by an organism is the product of three terms that determine the survival probability of the organism's genome, integrated over however many bouts of reproduction it engages in over its lifetime: the number of offspring produced (n_o), the expected longevity of those offspring (l_o), and the coefficient of genetic relationship between the parent and its offspring (r_{po}):

$$f = f_b(n_o)(l_o)(r_{po})$$

The function *f* represents the "fitness", or copying success of the entire genome over a specified period of time (such as the entire lifespan if lifetime fitness is desired), but also gives us the number of copies of any individual gene (technically, *allele*) surviving to any given point in time at which offspring survivorship is assessed. This is because the coefficient of genetic relationship (r_{po}) is equal to the probability of any given offspring carrying that particular gene, by the application of Bernoulli's Theorem. The probability of that particular gene surviving (r_{po}) is then multiplied by the number of offspring being produced (n_o) and the probability of those offspring surviving to a certain point in time (l_o). It is necessary to include the term (r_{po}) as a variable in this expression because it is not always equal to 0.50, even in sexually-reproducing diploid species, especially under conditions of assortative mating. This product represents the instantaneous replication of a genome achieved by any single bout of reproduction, and the integral over time sums those instantaneous products over the lifespan (for a complete mathematical derivation of this formulation, see Figueredo & Wolf, 2009; Wolf & Figueredo, 2011).

The second 'term of art' that needs clarification is *fertility*, as opposed to *fecundity*, owing to the fact that they are used in the biological sciences in a more specialized way than in general discourse. Fertility is defined as the per capita rate of the number of children actually born to each woman (typically operationalized as number of children per 1000 women) in the population. Sometimes, the denominator of this expression is limited to a certain subgroup of women. For example, the *general* fertility rate includes the number of children born to women between the ages of 15 and 44; the *age-specific* fertility rate includes those born to women of a given age cohort (e.g., 25 years old); the *completed* fertility rate includes those born to women of a given cohort until the end of their reproductive careers (e.g., 44 years old). The distinction that causes the most confusion, however, is not between the subset of women being considered, but between the concepts of *fertility* and *fecundity*, which are commonly used interchangeably. Fecundity, however, is technically defined as a woman's *potential* or *ability* to produce children, not her actual *reproductive output*, also known as *reproductive success*. Thus, a woman using contraceptives may have a high fecundity but a low fertility. Therefore, when we observe that higher-IQ women have lower fertility than lower-IQ women, for whatever reason, we do not mean that they lack the capacity to reproduce, but only that they generally do so at a lower rate.

1.1 Dysgenesis and Eugenesis Redefined

In considering the concepts of dysgenics and eugenics, we contend that new definitions are needed because the traditional usage borders on the logically incoherent and quite commonly crosses the line into becoming technically misleading. As was mentioned in the previous section, dysgenics is typically described in terms of the potential for 'socially desirable' traits (such as intelligence, but also health and desirable personality or 'good character'), the carriers of which are frequently described as being 'more fit' (e.g. Lynn, 2001; Nyborg, 2012) to become scarcer in a population over time owing to differential patterns of fertility disfavoring these traits. The outcome of dysgenic fertility is sometimes also termed *dysgenesis* (Herrnstein & Murray, 1994). The antonym of dysgenesis is of course *eugenesis*, which describes the increased prevalence of the 'fitter' carriers of 'socially desirable' traits entailed by the presence of either eugenic natural or artificial selection pressures (i.e. due to the presence of a *eugenics program*). From the perspective of evolutionary theory, however, it presents something of a logical paradox to assert that phenotypes that are 'less fit' are systematically out-reproducing those that are 'more fit'. This construction appears to fly in the face of the fundamental biological definition of Darwinian fitness, because the carriers of these presumably 'undesirable' traits are generally described as increasing their numbers at the expense of those with the presumably 'desirable' ones. For that reason, among others, some have concluded that what constitutes 'social desirability' is no more than a cultural construction, meaning that it represents an arbitrary value judgment by specific groups or individuals, and that these characterizations might be biased in a self-serving manner, consistent with their social, economic, or reproductive interests (Mackintosh, 2002).

To be able to employ the words 'dysgenic', 'eugenic' and related terms more productively in scientific discourse (if one wishes to retain them at all, given that the words imply 'undesirable' and 'desirable' outcomes respectively), we therefore need to provide a more objective reframing based on the evolutionary ramifications of the phenomena. We propose that this can be accomplished by viewing these processes within the framework of a multi-level selection model, which conceives of a 'dysgenic trait' as one that has the potential to confer a competitive advantage or *benefit* at an individual level, but imposes a competitive disadvantage or *cost* at the group level. In this multilevel selection model, 'dysgenic fertility' may thus occur when individuals bearing such traits are favored in within-group competition by individual-level selection, but the social groups to which they belong are disfavored in between-group competition as a consequence in group-level selection (Cattell, 1972; Figueredo, 2012). Conversely a 'eugenic trait' is one that has the potential to confer a competitive advantage at the group level, while not necessarily conferring an advantage at the individual level, and possibly even producing a competitive *disadvantage* in individual selection (Figueredo, 2012; Hamilton, 2000). Such traits might include 'heroic' levels of altruism during a time of war, where the principal beneficiary is the group rather than the individual.

The multi-level selection model dates back to Charles Darwin (1871), furthermore eugenicists in the early 20th century were clearly cognisant of the implications of multi-level selection for the proliferation or reduction of traits deemed to be 'socially desirable' (e.g. Krischel, 2012). A major line of supporting evidence for this comes from the documentary historical evidence indicating that during the heyday of the eugenics movement in the 1900's to the 1930's and even more recently, many of its most forceful advocates were politically *collectivistic*, irrespective of whether they believed in eugenics as a means of bettering a nation or ethnic grouping, or whether they were universalists, who believed that eugenic measures had to be adopted by mankind as a whole (Glad, 2006). Implicit, and sometimes also explicit in their pronouncements was the idea that eugenic measures must serve the collective good, if the collectivity is to prosper in the long run. A classic universalist eugenics advocacy document is the *Social Biology and Population Improvement* document (commonly known thereafter as the *Eugenicist's Manifesto*), which was produced by a group of eminent British and American biologists and geneticists (including Darlington, Haldane, Huxley, Dobzhansky, Muller, Price and Waddington amongst others) under the group-name of *Science Service* (1939). These biologists were explicit in their collectivism, stating for example that "some effective sort of federation of the whole world, based on the common interests of all its peoples" (p. 521) will be necessary for ameliorating the "economic and political conditions which foster antagonism between different peoples, nations and 'races'" (p. 521; quotes in original), which they in turn saw as a threat to the adoption of eugenic measures. Another potential hindrance to their eugenical aims were the economic social and economic conditions having the potential to impose opportunity costs on high-ability women in terms of childrearing capacity. These could only be alleviated via "an organization of production primarily for the benefit of consumer and worker" (p. 521), which would in turn necessitate that "dwellings, towns and community services generally are reshaped with the good of children as one of their main objectives" (p. 521). Even the suite of traits considered by these

8

eugenicists to be socially desirable include a kind of social or group mindedness, or "temperamental qualities which favour fellow-feeling and social behaviour rather than those... which make for personal 'success', as success is usually understood at present" (p. 521). Another, relatively more recent and also more explicit example of collectivist eugenic advocacy can be seen in the writings of Raymond B. Cattell (1972, 1987) on the topic of *Beyondism* – an ethical-religious system designed to promote eugenical aims and spread eugenical virtues via the encouragement of competition between different ethno-cultural groups. Cattell (1972) was the first to articulate a multi-level selection basis for thinking about the distinction between eugenic and dysgenic traits using modern evolutionary terminology, illustrating it thusly:

"For the processes of within-group individual selection and between-group selection of social organisms are not merely potentially independent, producing different results, but especially in regard to such vital traits as superego strength and self-sacrificing tendencies systematically undoing each other. Certainly one can see that it is easily possible for the within-group selection of individuals, i.e., the relative survival rates among individuals, to produce genetic types and tendencies to behavioral habits highly favorable to selfish individual survival but in the end incompatible with the survival of the group." (p. 84).

In a subsequent work Cattell (1987) illustrates his belief in the need for collective agency in the realization of group-level eugenic outcomes, stating that:

"A group positively planning well for its future will employ all three of the above: (1) differential birth/death rates, (2) rhythms of segregation and well-chosen hybridization, and (3) creation of mutations along with genetic engineering… These methods we need to use toward group goals to bring about by a collective movement of its citizens (a) *survival* of the group, and (b) launching out on its own evolutionary adventure" (p. 210-211, italics in original).

Consistent with the contention that eugenic fertility is favoured under conditions of inter-group conflict, where group selection acts to promote cultures in which socially altruistic individuals are fairly prevalent, these collectivist eugenicists believed that it was necessary to increase the frequency of group selected altruistic phenotypes at the expense of individually selected ones. Even though the signatories of the *Eugenicist's Manifesto* were universalists in that they clearly considered mankind to be their unit of collectivity, and were explicit in their desire to see conflict eliminated, the practice of collectivism in the 20th century, even amongst those regimes with comparable ideology to some of these biologist eugenicists (i.e. the Soviet Union), involved considerable inter-group conflict in reality. Amongst these universalist eugenicists there may even have been an implicit awareness of this, hence their emphasis on the social desirability of the sort of self-sacrificing behaviours that would benefit the group under such conditions. Not all eugenicists were so coy in explicating a preference for inter-group competition, Cattell (1972), a universalist eugenicist himself, evidently strongly believed in the desirability of such competition for the furtherance of eugenical aims, advocating what he termed 'cooperative competition', which is the view that:

"like players in some greater more vital game than men usually play, cultural groups recognize that the maintenance of inter-group competition is indispensable to evolution and they agree to cooperate in whatever rules are necessary to maintain it in effective action" (p. 86).

The definitions of dysgenesis and eugensis that we favor here are in principle measurable and quantifiable, and do not depend on culturally-relative value judgments or self-serving biases. For example, they do not take any *a priori* position as to whether individual-level or group-level interests should be considered to have primacy over the other. The classic view in multilevel selection theory is that the evolutionary outcome depends on the relative strengths of the selective pressures (D. S. Wilson, 2002; Nowak, Tarnita & E. O. Wilson, 2010), much the same way that the contrary pressures of natural and sexual selection are presumed to operate on the optimal length of a peacock's tail: too short a tail confers a competitive disadvantage in mating whereas too long a tail confers an aerodynamic disadvantage in predator-avoidance. No rational person has taken a strong emotional or ideological stance on which of these contrary selective pressures either the male or female peacocks should be preferentially attending to on abstract ethical, political, or moral grounds. It is important to note that in this important respect we depart radically from the eugenicists of the past who either implicitly or explicitly ascribed *differential normative value* to the adaptive value of traits with respect to these two levels of selection, typically favoring group survival and reproductive interests over individual ones as their criteria for the regime of artificial selection envisioned. These early eugenicists thus favored *artificial* selection over natural because they believed the latter was shaping our species in the wrong direction, and did not recognize explicitly that the same traits might have different adaptive values when considered with respect to different levels of natural selection (although Darwin explicitly did so as far back as 1871 in *The Descent of Man*).

Although as citizens in a free society, we and others have the right to espouse any normative position we choose, as scientists we take no such normative position and merely describe the observable historical effects of these two opposing selective pressures on the evolution of human nature.

In summary then, the fitness reduction produced by 'socially undesirable' traits therefore has to be thought of in terms of group selection, where the aggregate efficiency of a competing group is diminished by the differential reproductive success of those with lower levels of 'socially desirable traits', which renders dysgenesis genuinely maladaptive at the level of competing groups (and in terms of group fitness) rather than at the level of individuals. This makes sense if one considers carefully the traits that are considered to be 'socially desirable'. Individuals are usually not the recipients of widespread social approval by virtue of their extraordinary efforts to promote their own self-interest. Instead, individuals are motivated by what Darwin (1871) called the 'praise and blame of his fellows' for self-sacrificing, altruistic behaviors that benefit their social group.

In this monograph, we will explicitly address these theoretical concerns after reviewing some of the empirical evidence that is directly relevant to the dysgenesis hypothesis as it pertains to the reproduction of IQ.

1.2 Differential Fertility by IQ

The first half of the 20th century saw tremendous interest in the measurement of the extent of dysgenic trends for IQ in Western populations. This was achieved by documenting the negative relation between patterns of fertility, sibling numbers and scores on intelligence tests (e.g. Bradford, 1925; Burks & Jones, 1935; Burt, 1948; Cattell, 1936, 1937; Chapman & Wiggins, 1925; Damrin, 1949; Giles-Bernardeli, 1950; Lentz, 1927; Moshinsky, 1939; Nisbet, 1958; Papavassiliou, 1954; Roberts, Norman & Griffiths, 1938; Sutherland, 1930; Sutherland & Thomson, 1926; Thomson, 1946, 1949; Thurstone & Jenkins, 1931; P. E. Vernon, 1951). However early attempts to quantify the long-term negative impact on IQ of these relations using cross-sectional studies found the opposite effect – namely that IQ had been rising instead of falling (Burt, 1948; Cattell, 1951; Emmett, 1950; Maxwell, 1954; Smith, 1942; Thomson, 1949; Tuddenham, 1948). This became known as "Cattell's paradox" (Higgins, E. Reed & S. Reed, 1962). What these early researchers had uncovered was evidence for what are currently termed secular gains in IQ, which are known to be have been occurring since at least the 30's in the West at a rate of approximately 3 points a decade (Flynn, 1987, 2009a). Despite this, the preponderance of studies on the fertility-to-IQ relation reveal consistent negative correlations between the two across time (Lynn, 1996, 2011; Lynn & van Court, 2004; Nyborg, 2012; van Court & Bean, 1985).

Again, the technical terminology used within intelligence research may be confusing to the non-specialist. To ameliorate this, we present a breakdown of the different complementary theoretical terms that presumably add up to an observed IQ score in Table 1. The contrast between the additive increments in IQ attributable to genetic and to environmental influences is well known, however the general and specific components of variance in IQ are widely misunderstood.

Spearman's 'g' is defined as the unitary general factor representing the shared variance among a 'positive manifold' of correlated tests of distinct mental abilities (Bartholomew, 2004; Spearman, 1904, 1927; Jensen, 1998). This is the component that

Spearman became most famous for proposing. However, Spearman (1927) referred to his theory as a 'Two Factor' rather than a 'One Factor' theory, because he also highlighted the importance of the specific ('*s*') factors representing the more specialized components of the total IQ score. To illustrate this, it is well known that there exist seemingly distinct cognitive abilities, such as visuospatial ability and verbal ability. Spearman's two factor model posits that the correlation amongst these seemingly distinct abilities constitutes the variance attributable to general intelligence or '*g*', whereas the non correlated variances are specific to each ability and constitute '*s*'.

Table 1. A Cross-Tabulation of the Complementary Components of Variance Involved in Human Intelligence

Terms of Art	Genetic Influence	Environmental Influence	Sums
General Factor	Heritable '*g*' (*g.h*)	Environmental '*g*' (*g.e*)	'*IQ.g*'
Specific Factors	Heritable '*s*' (*s.h*)	Environmental '*s*' (*s.e*)	'*IQ.s*'
Sums	Heritable '*IQ*' (*IQ.h*)	Environmental '*IQ*' (*IQ.e*)	Phenotypic IQ ('*IQ.p*')

Both '*g*' and the various sources of '*s*' then each have their corresponding genetic and environmental sources of influence, leading to the breakdown shown in Table 1. Although psychometric estimates of the 'common factor' score correspond more closely to the theoretical definition of *g*, an IQ score is instead interpretable as an *aggregate* of all these theoretical components of variance, and not just the common factor variance component. These components are not directly observable, but are inferential constructs based on many years of psychometric research.

The proportion of the variance in IQ accounted for by the *g*-factor (denoting what in psychometrics is known as the *common factor* variance) has not been constant over history, with a number of studies indicating that it has declined (Juan-Espinosa, Cuevas, Escorial, & García, 2006; Kane, 2000; Kane & Oakland, 2000; Lynn & Cooper, 1993, 1994; O. Must, A. Must & Raudik, 2003a; O. Must, te Nijenhuis, A. Must, & van Vianen, 2009; Sundet, Barlaug & Torjussen, 2004; Woodley & Madison, in press) but today the factor accounts for about 50% of the variance (Brand 1996). Studies typically place the adult heritability of IQ at $\geq$75% (Bouchard Jr, 2004; Gottfredson, 1997; Neisser et al. 1996), with very similar estimates for the heritability of the *g* component (W. Johnson et al., 2007; Trzaskowski, Yang, Visscher & Plomin, in press), and lower estimates being typical for specific cognitive abilities (Carroll, 1993; W. Johnson et al., 2007).

As this is such a substantial proportion of the aggregate, differential fertility disfavoring intelligence should therefore act to reduce the population mean IQ with time. On this basis, a number of researchers (e.g. Loehlin, 1997; Lynn, 1996, 1999, 2011; Lynn & van Court, 2004; Meisenberg, 2010; Meisenberg & Kaul, 2010; Nyborg, 2012; Retherford & Sewell, 1988; Vining, 1982, 1995) have attempted to use the size of the 'selection differential' on IQ, measured using either the gradient of the fertility/IQ differential amongst those with completed fertility, or the IQ/family size correlation to estimate the *theoretical* decline in absolute levels of the heritable component of intelligence ('*IQ.h*') by multiplying the inferred decline in phenotypic aggregate intelligence (*IQ.p*) by the additive or 'narrow sense' heritability of IQ (Plomin, De Fries & McClearn, 1990).

Table 2: UK decline estimates based on Lynn (1996, 2011) for cohorts born between 1890 and 1955 (encompassing a 90-year period spanning 1890 to 1980).

Cohort birth year and interval	Predicted Declines in Phenotypic Intelligence (IQ.p) Based on Fertility Differentials	Estimated Declines in Genotypic IQ (per decade)
1890 (1890-1914)	-2.0	-1.64 (-0.68)
1915 (1915-1939)	-2.0	-1.64 (-0.68)
1940 (1940-1954)	-1.4	-1.15 (-0.82)
1955 (1955-1980)	-0.8	-0.66 (-0.26)
Total decline	-6.2	-5.08 (-0.56)

These researchers have produced theoretical decline estimates (using different heritability estimates of IQ) ranging from a low end of -0.12 Genotypic IQ points per decade (Retherford & Sewell, 1988) to a high end of -0.69 Genotypic IQ points per decade (Nyborg, 2012). Lynn (1996, 2011) computed an aggregate genotypic decline of -0.26 points per decade for British cohorts born post-1955. The mean of all modern (post-1980) decline estimates has been found to be very similar at -0.37 IQ points per decade (Rindermann & Thompson, 2011a). Earlier studies employed a different methodology in calculating IQ declines, i.e. they differentially weighted the parental and offspring IQ means based on indicators of fertility (typically sibling numbers), and used the difference between these means as a proxy for dysgenic phenotypic IQ (*IQ.p*) decline. Lynn (1996, 2011) computes a Genotypic IQ decline of -0.68 points per decade for British cohorts born in the late 19th and early 20th centuries (1890 and 1915 respectively) by averaging across the decline estimates produced in the studies of Burt (1948), Catell (1936), Sutherland

(1930) and Thomson, (1946). Studies in roughly the same period from the States (Lentz, 1927) suggest the *IQ.p* was declining at a rate of approximately 1.3 points per decade, which equates to a decline of as much as 1.1 points of Genotypic IQ per decade, given Lynn's average additive heritability estimate for IQ of 0.82. The disparity between estimates generated in the opening decades of the 20th century and those generated in later decades suggests the existence of a secular weakening trend in the intensity of dysgenic pressures. Lynn has found evidence for such a trend by comparing the size of the estimated dysgenic decline rates for four UK birth cohorts over a 90-year period (1890-1980).

These data are presented in Table 2 along with decadal Genotypic IQ decline estimates based on Lynn's preferred estimate of the additive heritability for IQ of 0.82.

Lynn has argued that this secular trend has resulted from fertility control technologies (contraceptives) being used overwhelmingly by members of the cognitive elite in the early decades of the 20th century. During the later decades contraceptives were used much more widely, hence the impact of dysgenic fertility would have been attenuated.

As was mentioned previously, however, attempts to directly measure dysgenesis have typically found the opposite effect, namely a large secular increase in IQ scores over time. This effect was first brought to wide-scale public attention by James Flynn (1987) and Richard Lynn (1983), and was subsequently termed the 'Flynn effect' (Herrnstein & Murray, 1994), or the 'Lynn-Flynn effect' (Rushton, 1999) in recognition of its co-promoter. Many theories have been proposed to account for the effect, most of which revolve around the role of 'environmental multipliers', which are improvements in environmental conditions (such as increased access to education, better nutrition, improvements in hygiene, lower exposure to neurotoxins, greater cognitive stimulation etc.) that have either independently or collectively leveraged large-scale gains in measured IQ since the opening decades of the 20th century (Dickens & Flynn, 2001; Eppig, Fincher & Thornhill, 2010; Lynn, 1989; Meisenberg, 2003; Neisser, 1997; Nevin, 2000; Woodley, 2012b).

1.3 Genotypic IQ, Eminence, and Significant Innovations

Even though dysgenesis on heritable intelligence (*IQ.h*) or Genotypic IQ remains largely theoretical owing to the confounding influence of the Flynn effect on attempts at direct measurement (Lynn, 1996, 2011), two studies present evidence that dysgenic fertility may have had real world corollaries. The first is Charles Murray's (2003) *Human Accomplishment* in which it was found that the per capita number of 'eminent' figures in science and technology has been declining since the late 1800's, whereas in previous centuries the number had been increasing. Utilizing Galton's (1869) method of 'convergent bibliography', Murray determined the relative eminence of individuals on the basis of their prominence across encyclopaedias and other reference works. While Murray does not attribute this trend to dysgensis (he argues instead that it resulted from scientists turning away from truth-seeking orientations associated with religious practices such as Thomism) it is nonetheless compatible with the idea that *IQ.h* has been declining since the mid 19th century, as some models argue that eminence or extraordinary talent results from possibly 'emergenic' (Jensen, 1997) combinations of high cognitive ability and specific conative (personality) traits (such as Psychoticism, which relates to creativity and unconventional thinking; Eysenck, 1995; Simonton, 1999). These models are multiplicative in that eminence and extreme talent can only emerge with precisely the right (rare) trait combinations, hence a decline in just one of the components of eminence is all that is needed in order to reduce the frequency of eminent individuals over time. It is also worth noting that as a stand-alone predictor of success in life, even at extraordinarily high levels, studies indicate that relatively small IQ differences matter in terms of career success in the sciences (Arneson, Sackett & Beatty, 2011; Robertson, Smeets, Lubinski & Benbow, 2010).

The second (Woodley, 2012a) involved a temporal multivariate analysis in which the relationship between a reconstructed historical trend in Western Genotypic IQ means and the per capita rates of significant innovation in science and technology (i.e. the numbers of significant events per year per billion of the world's population; Huebner, 2005) was investigated. Significant innovations, like eminent individuals, are defined in terms of their prominence across encyclopaedias and other reference works, and as with the eminent individuals who generate them, the data indicate that these are primarily the product of Western populations (Murray, 2003). These 'significant innovations' therefore constitute unambiguously high-impact or 'macro-level' developments, such as the discovery of calculus, the theory of evolution or putting a man on the moon. This is in contrast with the much more commonplace incremental 'micro-level' innovation, examples of which might include the production of a new variant on an existing design of computer or mobile phone. Innovation inventories principally strive to measure the former rather than the latter (Huebner, 2005). These were predicted to be especially sensitive to changes in Genotypic IQ means on the basis of Cattell's (1937) prediction that "Considerably more intelligence is required to discover scientific principals than is needed to use them once discovered." (p. 81).

In Woodley's (2012a) study, Genotypic IQ means were computed at decadal intervals between 1455 and 2005 for Western societies. A computer-generated model developed and published by Hart (2007, p. 124) was used to estimate the means for Genotypic IQ in Western populations living in the 15th century. The model assumes that the Genotypic IQ increased as a linear function of environmental harshness over time. Even though this is a simplified model, the simulated modern day IQs closely approximate those actually observed, hence a case could be made for the historical estimates being crude but reasonable proxies.

Between the 15th and 19th centuries, Woodley assumed that Genotypic IQ had been increasing in Western societies owing to the existence of positive correlations then prevalent between indicators of social status and fertility during this period (Clark, 2007; Skirbekk, 2008). From the 1800's onwards, this correlation became negative (Skirbekk, 2008), and it was at this point that Western populations transitioned into dysgenic fertility (Lynn, 1996, 2011). Between 1850 and 2005 two different estimates were used for the dysgenesis rate. One was based on the recently published 'decay model' of Nyborg (2012), which predicts Genotypic IQ declines of around -0.69 IQ points per decade, with a total loss of around 9.0 IQ points between 1850 and the present. The second estimate employed a more conservative decline rate based on the assumption of a lower dysgenesis rate per decade (around -0.35 points per decade). This yielded a total loss of about 5.0 IQ points since 1850.

These Genotypic IQ measures were used along with historical measures of homicide and literacy rates, per capita Gross Domestic Product, measured in terms of adjusted Purchasing Power Parity and estimates of the Flynn effect based on the assumption of sigmoidal growth with the biggest gains occurring between 1900 and 2000 (Crepin, 2009; Meisenberg, Lawless, Lambert & Newton, 2005) as predictors of the per capita significant innovation rate measure, which shows substantial declines starting in the latter half of the 19th century and continuing into the present (Huebner, 2005). It was found that the Genotypic IQ estimates were the best predictors of changes in the per capita rate of innovation throughout the period studied, and that this relation was robust to the incorporation of basic controls for temporal autocorrelation. Furthermore the Flynn effect had only a very small positive influence on innovation rates according to the path analysis. This suggests that the gain in measured IQ is not substantially affecting a civilization's capacity for innovation. The strongest correlate of the Flynn effect was GDP (PPP) per capita, and in multiple regression, an aggregate of these was most strongly predicted by an aggregate of literacy and homicide rates, which is consistent with recent research into the social benefits of diminishing violence (Pinker, 2011).

A key prediction made in Woodley (2012a) is that changes in the size of the 'smart fraction' will strongly mediate the impact of mean changes in level of Genotypic IQ on significant innovation rates. *Smart Fraction Theory* holds that it is this fraction of the population, rather than the average, that is most important for progress in science and technology (La Griffe du Lion, 2004). Furthermore, given the fact that 'smart fractions' constitute the high-IQ 'slice' of a population, with different researchers assigning different cutoffs (Rindermann, Sailer & Thompson, 2009), their size is highly sensitive to the population mean IQ owing to the fact that IQ is distributed in a normal fashion, and a small shift in mean IQ can have a much larger impact on the tails of the distribution. Herrnstein and Murray (1994) illustrate this by pointing out that a 'mere' three-point decline in the mean IQ of a population (from 100 to 97) would reduce the size of the fraction of the population containing people with IQs ≥ 135 by 42%.

Several predictions from smart fraction theory have already been validated. Rindermann, Sailer and Thompson (2009) found that national smart fraction size was a stronger predictor of indicators such as patent rates, scientific papers published, Nobel prizes won etc., than was the population mean IQ. Another study (Rindermann & Thompson, 2011b) found strong correlations between the size of a nation's 95th cognitive ability percentile and a population size weighted index of national productivity in terms of the numbers of eminent individuals in science and technology, derived from Murray's index. This further reinforces the idea that historical variation in the frequencies of Murray's eminent individuals may be related to the effects of eugenic and dysgenic fertility on the smart fractions of populations.

1.4 Convergent Evidence: Relations among Dysgenic fertility, Flynn Effects and General vs. Specific Factors of Intelligence

Changing innovation rates and per capita numbers of eminent individuals are not the only evidence that Western populations are currently in the dysgenic phase of a population 'super-cycle' in eugenic and dysgenic fertility for heritable intelligence, or *IQ.h* (Cattell, 1987; Meisenberg, 2007; Weiss, 2007). For example, another significant line of evidence for dysgenesis (illustrated in figure 1) is the observation that the latency of simple reaction times, which share common genetic variance with *g* (Rijsdijk, P. A. Vernon & Boomsma, 1998), have increased substantially since Galton first started measuring them in the 1880s (Silverman, 2010).

Another important converging line of evidence comes from studies directly investigating the affinities between dysgenic fertility, Flynn effects and *g*. An important tool for determining the degree to which a source of individual or group differences in IQ exhibits an affinity for *g* is the method of correlated vectors

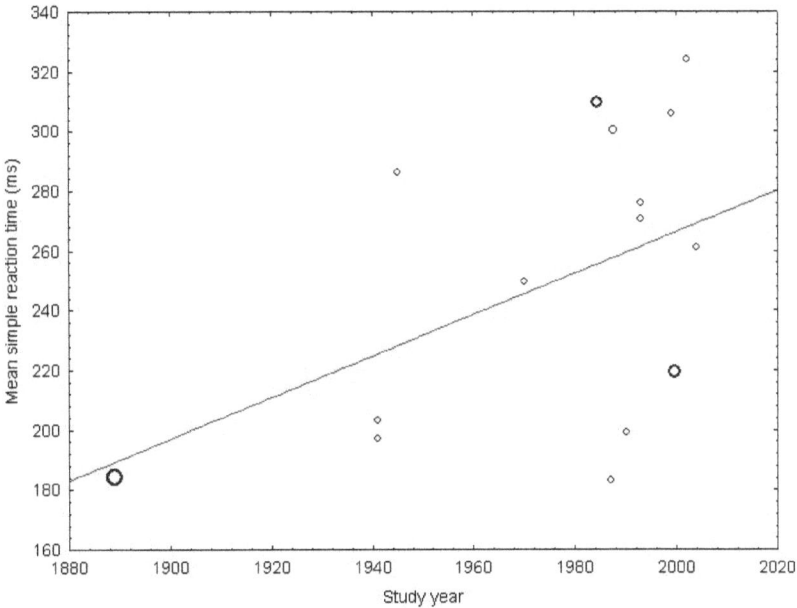

Figure 1. Simple reaction time means vs. year for *N*=15 effects. Data from Silverman (2010). Male and Female reaction time means have been combined on a weighted basis where data on both sexes are available. The bubble-size indicates the sample size. The unweighted temporal correlation is *r*=0.505 and is borderline significant at *P*≈0.05 for an *N* of 15 studies.

(Jensen, 1998). This method involves the correlation of the rank of the magnitude of a given association between IQ and a source of individual or group differences with the rank of the *g*-loadings of the tests on which the association is established.

Studies involving the US population representative National Longitudinal Survey of Youth (NLSY) reveal that the negative correlation between fertility and IQ (the gradient of dysgenic fertility) is most strongly pronounced on the *g* factor rather than on the relatively more *s*-loaded subtests, when examined individually (Meisenberg, 2010; Meisenberg & Kaul, 2010). As illustrated in figure 2, at the level of the whole NLSY population,

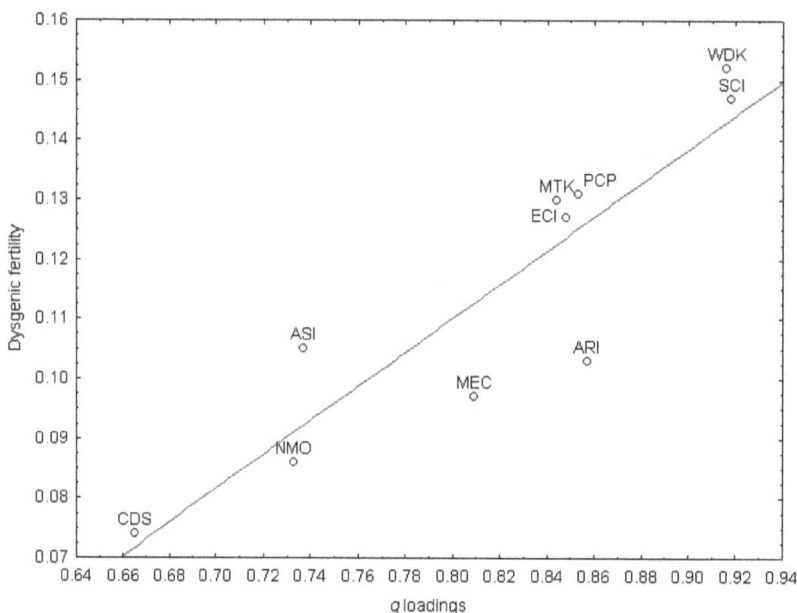

Figure 2. Scatter plot illustrating the positive relationship between dysgenic fertility (the correlation between IQ and fertility; rescaled from negative to positive so as to work in terms of effect magnitudes) and the *g*-loading of the 10 Armed Services Vocational Aptitude Battery (ASVAB) subtests on which the dysgenic fertility effects were measured, for the whole of the NLSY'79 sample exhibiting completed or nearly completed fertility (*N*=8,110 individuals). The vector correlation (*r*) is 0.899, and is significant when Spearman's rank order correlation is used to determine significance with an *N* of 10 subtests. The subtests are keyed as follows: SCI=Science; ARI=Arithmetic, WDK=Word Knowledge, PCP=Paragraph Comprehension, NMO=Numerical Operations, CDS=Coding Speed, ASI=Auto-Shop Information, MTK=Math Knowledge, MEC=Mechanical Comprehension, ECI=Electronic Information. Based on the analysis of Woodley and Meisenberg (in press).

the vector correlation between dysgenic fertility magnitudes and the *g*-loadings of the 10 Armed Services Vocational Aptitude Battery (ASVAB) subtests is nearly monotonic at 0.899 for an *N* of

8,110 individuals, and is statistically significant when this parameter is calculated using the more stringent Spearman's rank order correlation (Woodley & Meisenberg, in press).

This indicates that dysgenesis when measured using the size of the negative correlation between IQ and fertility, exhibits a very strong affinity with g. Studies examining the relation between Flynn effect gains and the rank of test g-loadings typically find negative correlations between the two by contrast (Jensen, 1998; O. Must, A. Must & Raudik, 2003a, b; Rushton, 1999; Rushton & Jensen, 2010; te Nijenhuis, in press; te Nijenhuis & van der Flier, 2007, in press). A psychometric meta-analysis of the results of 11 studies (N=16,663 individuals) reporting both Flynn effects and subtest g-loadings found that the vectors correlated significantly at -0.38 (te Nijenhuis & van der Flier, in press), indicating that Flynn effect gains are typically most pronounced on the subtests that are least g-loaded. Rushton and Jensen (2010) have argued that this could be an underestimate of the true independence of the Flynn effect from g, which may be closer to -1.00 when $g.e.$ is disentangled from $s.e.$ Perhaps the strongest evidence for the Flynn effect's independence from g are studies investigating whether or not the effect is associated with the property of factorial or measurement invariance. If the factor on which the Flynn effect occurs at time point a is the same as the factor on which it occurs at time point b, then the two factors should be isomorphic. All studies investigating this issue thus far have found that the Flynn effect is not associated with factorial invariance at the level of g, which means that it is not associated with changes in the amount of g (O. Must, te Nijenhuis, A. Must, & van Vianen, 2009; Wicherts et al., 2004). Even at the level of individual Raven's Matrices items, the effect is not associated with measurement invariance (Fox & Mitchum, in press). This finding compliments the aforementioned observation that the Flynn effect appears to be associated with a 'secular weakening' of g over time, as this process is evidence that IQ tests are increasingly measuring performance on s factors rather than g with the passage of time.

These three findings combine to suggest that secular gains in measured IQ are primarily associated a) with heterogeneous gains in narrow abilities and b) that these abilities are also substantively 'hollow' with respect to g, and are furthermore becoming 'hollower' with time (i.e. less strongly correlated with g).

As subtest heritabilities increase monotonically (i.e. the vectors correlate perfectly) with their g-loadings (van Bloois, Geutjes, te Nijenhuis & de Pater, 2009), dysgenic fertility is therefore maximally pronounced on the subtests that are most strongly influenced by heritable factors, which suggests that it is primarily a genetic effect, whereas the Flynn effect is greatest on subtests that are relatively less heritable and are therefore theoretically more responsive to environmental improvements, such as better nutrition, education and cognitive training.

Based on Table 1, and consideration of the results of applying the method of correlated vectors to dysgenic fertility and Flynn effects, a potential solution to Cattell's paradox presents itself; this is based on the observation that dysgenesis primarily occurs on $g.h$, whereas the Flynn effect primarily occurs on $s.e$. As these sources of variance are free to change independently of one another, dysgenesis and the Flynn effect can co-occur and are therefore not necessarily incompatible effects. This model could be termed the 'co-occurrence model', and represents a substantial theoretical departure from the currently most popular solution to Cattell's paradox, based on what could be termed the 'attenuation model', which holds that massive phenotypic IQ gains via the Flynn effect are masking relatively much smaller losses in Genotypic intelligence due to dysgenesis (Burt, 1948; Lynn, 1996). Loehlin (1997) illustrated this model succinctly with the analogy of rising tides (representing the Flynn effect) and leaky boats (representing dysgenesis). It is important to note that these models make contrasting predictions about the social impacts of dysgenesis with the former predicting a historical impact stemming from dysgenesis on *g.h*, which has not been offset by massive gains in *s.e*, whereas the latter predicts that owing to a potential involvement of g in secular gains, dysgenic effects will only start to have an impact when the Flynn effect ceases. A key prediction of the attenuation model, falsified by the observation of declining innovation rates and rates of eminent individuals, is that the Flynn effect has not just masked dysgenesis at the level of test performance but in actuality also.

This finding also has implications for the aforementioned increase in simple reaction time latency because the association between IQ and reaction times, like dysgenic fertility magnitude is substantially more pronounced on tests exhibiting higher *g*-loadings (Jensen, 1998, 2006), which suggests a possible causal connection as information processing speed measures might constitute an especially good indicator of *g*, in that they might both relate to common fundamental neurophysiological processes such as the periodicity of the central nervous system (Jensen, 1998, 2006, 2011), and might thus be highly sensitive to dysgenesis. The idea that reaction time constitutes an especially 'pure' measure of *g* also is bolstered by the observation that in the absence of a Flynn effect it is potentially measurement invariant with respect to *g*. This interpretation is supported by the absence of Flynn effects on other information processing speed measures also, such as inspection time (Nettlebeck & C. Wilson, 2004).

An important line of supporting evidence for this clustered effects interpretation comes from a study by Rushton (1999), who identified a positive manifold amongst a variety of different measures of two genetic and part-genetic variables, which exhibited positive vector correlations with respect to subtest *g*-loadings. These variables included Black-White cognitive ability differences and inbreeding depression scores. Rushton (1998) proposed that such effects be termed Jensen effects, in honor of Arthur Jensen, who first proposed the method of correlated vectors (Jensen, 1998). It was also found that the Flynn effects in the sample loaded preferentially on a separate principal component, suggesting that they were largely causally distinct from the genetic and part-genetic variable cluster.

The range of individual and group differences variables found to be associated with the Jensen effect are quite broad, and in addition to the aforementioned, include neurophysiological traits such as evoked potentials (Eysenck & Barrett, 1985; Schafer, 1985), glucose metabolic rate (Haier, Siegel, Tang, Abel, & Buchsbaum, 1992) and intercellular brain pH (Rae et al., 1996), in addition to heterosis (hybrid vigor) (Nagoshi & R. Johnson, 1986), brain size (Rushton & Ankney, 2009), fluctuating asymmetry (Prokosch, Yeo & Miller, 2005) and sex differences in cognitive ability (Nyborg, 2005).

Importantly, giftedness has also been found to be strongly associated with the Jensen effect, and hence with *g* (van Bloois, Geutjes, te Nijenhuis & de Pater, 2009). This is significant as a major supporting line of evidence for the dysgenesis hypothesis comes from the previously discussed observation that the profoundly gifted (i.e. eminent individuals) have been diminishing in number on a per capita basis (Murray, 2003). This decline has been occurring over roughly the same time period as the increase in latency of simple reaction times (Silverman, 2010) and the transition into dysgenic fertility in the West (i.e. the mid to late 19th century onwards; Lynn, 1996, 2011; Skirbekk, 2008). Like Rushton's (1999) genetic cluster, the phenomenological commonalities amongst these three Jensen effect related variables combine to define a nexus, which could be termed the *dysgenesis syndrome*.

Other effects have also been found to share phenomenology with the Flynn effect in terms of being anti-Jensen effects, such as the IQ gains accrued via retesting (te Nijenhuis, van Vianen & van der Flier, 2007), which reinforces the idea that the Flynn effect is principally environmental and non-g in origin.

It must also be noted that the findings of the Woodley (2012a) study indicate, consistent with the co-occurrence model, that mean changes in Genotypic IQ or heritable IQ (*IQ.h*) and the Flynn effect have had different impacts on aspects of the historical development of the West. Whereas changes in Genotypic IQ predict trends in per capita significant innovation rates, the growth in the Flynn effect strongly parallels the near logarithmic growth in GDP PPP per capita.

One possibility therefore is that as the Flynn effect is associated primarily with the proliferation of cognitive specialization (i.e. the development of narrow, highly s-loaded abilities), which has facilitated more fine grained divisions of labor and concomitant 'micro-level' innovation of a sort that would have increased the aggregate economic efficiency of populations via the operation of Ricardo's *Law of Comparative Advantage* (Ricardo, 1891; Woodley, 2011a, b, 2012a, b). Significant innovations stemming from higher population *g.h* may therefore have facilitated the sorts of initial environmental improvements necessary to catalyze the Flynn effect by the turn of the 20th century. This process would then have become self-catalyzing to the extent that subsequent declines in population *g.h* did not hinder either the growth in wealth or run-away environmental improvement. An implication of this is that there exist negative social multipliers in addition to the positive ones discussed previously. In other words, if *g.h* falls below some critical threshold then it might degrade environmental quality (i.e. via diminished educational standards) to the point where the Flynn effect ceases or even reverses itself (Meisenberg, 2003). Evidence that this might already have happened in some Western countries (e.g. Australia, Denmark, Norway, Sweden and the UK) comes from the finding of no further gains or even net losses in some instances on the sorts of culture-fair fluid-IQ (i.e. raw problem solving ability) measures that in previous decades gave rise to appreciable Flynn effects (Cotton et al., 2005; Emanuelsson, Reuterberg & Svensson, 1993; Flynn, 2009b; Shayer, Ginsburg & Coe, 2007; Shayer & Ginsburg, 2009; Sundet, Barlaug & Torjussen, 2004; Teasdale & Owen, 2005, 2008).

The idea that the interplay amongst *both* genetic and environmental factors might be driving these recently observed anti-Flynn effects lends its-self to the prediction that such effects should be mildly associated with the Jensen effect. A recently published study (Woodley & Meisenberg, 2013) has found evidence for just this amongst the 1950-1990 Netherlands birth cohort in a sample of 63 effect sizes exhibiting a combination of both Flynn and anti-Flynn effects. When considered separately, the anti-Flynn effect magnitudes exhibited a borderline statistically significant vector correlation of 0.406 with subtest g-loadings in contrast to the Flynn effects, whose magnitudes were not significantly associated with subtest g-loadings. This constitutes a potentially confirmatory test of the co-occurrence model, as IQ gains and declines in the same birth cohort on the same test batteries exhibit apparently distinct causal phenomenologies, with the latter, unlike the former exhibiting a significant affinity for g and therefore the nexus of the dysgenesis syndrome.

Finally, an apparent exception to the dysgenesis syndrome is the observation that endocranial volume and brain size (as indicated by head circumference) have been increasing in tandem with the Flynn effect (Lynn, 1989, 1990). Brain size is associated with the Jensen effect (Rushton & Ankney, 2009) hence it might be expected to trend negatively over time, consistent with the other components of the syndrome. Brain size exhibits an unusual combination of a high heritability and a low coefficient of additive genetic variance (suggestive of stabilizing rather than direction selection), furthermore it is genetically uncorrelated with fitness indicators known to share genetic variance with g which suggests a potential role for different selective processes in maintaining within-population variance in both traits (Miller & Penke, 2007). This model indicates that a change in brain size does not have to be accompanied by a change in g however this does not explain the apparent Flynn effect on brain size. Lynn (1990) attributes this change to improvements in nutrition, which have also boosted height. One possibility therefore is that the association between the Flynn effect and the gain in brain size is an artifact of increasing height. Consistent with this is Lynn's (1990) finding that the brain-size to IQ relationship effectively vanishes when height is controlled. Lynn interprets this as indicative of the *joint* effects of improved nutrition on height, phenotypic IQ and brain-size however. Another possibility is that the *gain* in brain-size is not itself a Jensen effect owing to its potential association with differential gains in non-verbal (such as visuospatial ability) rather than verbal abilities (Lynn, 1990). This would indicate that brain-size might have increased so as to accommodate the enhanced development of sources of $s.e$ which are demanding in terms of brain substrate rather than a hypothetical gain in $g.e$. This also suggests that cognitive endophenotypes such as brain size are poorer contemporary indexes of g owing to the Flynn effect, compared with the past.

1.5 Testable Predictions

In this monograph, we report the results of a critical test of the prediction from smart fraction theory that the impact of mean changes in *g.h* on innovation rates should be mediated by the numbers of 'eminent' figures in science and technology. We then evaluate our findings in the discussion in the context of multiple converging lines of evidence supportive of the multi-level selection model.

2. METHODS

2.1 Measures

The estimates used here of the Western heritable component of general intelligence, or *g.h* for short, incorporate the decadal decline estimates derived from Lynn's (1996, 2011) time-series (Table 2), and are thus sensitive to the secular trend in dysgenesis intensity. Lynn's UK dysgenesis estimates do not extend back to 1850; we therefore interpolate estimates for *g.h* for that year using the 1890 to 1914 dysgenesis rate. We furthermore make the assumption that these trends are paralleled throughout the West (e.g. Nyborg, 2012; Skirbekk, 2008). Whereas Lynn attempted to produce a Genotypic IQ estimate by multiplying the theoretical decline in IQ inferred from the magnitude of the selection gradient by an IQ heritability of 0.82; we utilized a slightly lower estimate of 0.75, which is more consistent with the preponderance of the literature (Gottfredson, 1997; Neisser et al., 1996). It must be noted that the Genotypic IQ estimates employed here do not correspond precisely to *g.h*. This is because IQ scores are an aggregate of variance proportions uniquely attributable to both the heritable and environmental components of *g* and *s* (and also error), as was demonstrated in Table 1. To estimate the fraction of the decline attributable uniquely to *g.h*, the component on which the effects of IQ dysgenesis should be maximally concentrated owing to the Jensen effect on dysgenic fertility, it would be necessary to incorporate an estimate of the variance proportion in IQ attributable to *g* into the model (for details of the mathematics behind this procedure see Appendix B). As was mentioned previously, this proportion seems to have historically decreased due to the Flynn effect over time by a currently unknown amount, hence we do not attempt to extract the precise fraction of the decline uniquely attributable to changes in the levels of the g. variance component from the $IQ.h$ or Genotypic IQ aggregate.

We propose therefore that the Genotypic IQ measure employed here be considered a *proxy* for declines in the $g.h$ component and will henceforth refer to our Genotypic IQ estimates as $g.h$. In support of this procedure, recent papers (W. Johnson et al., 2007; Trzaskowski, Yang, Visscher & Plomin, in press) have estimated the heritability of general intelligence ($g.h$) at between 0.73 and 0.77, which is nearly identical to the aggregate heritability of IQ ($IQ.h$) of 0.75 utilized here. It is therefore mathematically inevitable that the *aggregate* heritability of *all* specific abilities ($s.h$) will be approximately equal to that of general intelligence ($g.h$), although heritabilities and environmentalities may vary substantially among the *individual* specific abilities, and their cognitive constituents, thus providing a substrate upon which the Flynn effect can operate in the case of ultra-high environmentality sources of s. Hart's (2007) model was used to derive a proxy estimate of Western (north and south European) Genotypic IQ for the year 1455, as per Woodley (2012a). This estimate was considered reasonable on the grounds that it was lower than the estimate for 1850, hence is compatible with the observation of 'eugenesis' for IQ proxies during this period. One additional step was to multiply these 'eugenesis' estimates by a heritability value of 0.75 to harmonize them with the 'dysgenesis' estimates. Although this additional multiplication was probably unnecessary based on theoretical considerations, it was nonetheless needed to be able to transform the estimates based on Hart's model into the same metric as those based on Lynn's model. In either case, all such scores were standardized into z-scores for all of our further analyses, so that the absolute metric used (once rendered internally consistent) was irrelevant to our final results.

Western Flynn effect gains after 1900 were estimated on the basis that they have been observed to be fairly uniform over the last seven or so decades of the 20th century (three points per decade Flynn, 1987, 2009a). 'Pure' Flynn effects before 1900 can be assumed to have been weak in the prior centuries, owing to the widespread absence of the sorts of environmental improvements and multipliers believed to have facilitated them (i.e. generalized education, adequate nutrition, proper hygiene etc.). Crepin (2009) has argued that 50 probably represented the absolute lower limit pre-Flynn effect IQ for Western populations in the past; a point echoed in subsequent research, which is based on the idea that an IQ in the 50s represents the lower bound mean IQ for contemporary pre-modern populations (i.e. hunter-gatherers) that have yet to be exposed to Flynn effect inducing cognitive stimuli (Oesterdiekhoff, 2012). The problem with these assertions is that superficially, they make psychometrically untenable assertions about the nature of historical populations that are not supported by the data. For example, it has been noted that although people at the turn of the 20th century would have invariably scored substantively lower than modern people on certain psychometric IQ measures, they were nonetheless not functionally retarded relative to modern populations (Flynn, 2009a). One way around this interpretative problem is based on the observation that the Flynn effect is principally occurring on something largely unrelated to $g.h$ – i.e. sources of $s.e$ that adapt individuals to the demands of contemporary modernity, such as the capacity to think in empirical rather than practical terms about the relations amongst phenomena (Flynn, 2009a). Hence contemporary populations will be better able to understand contemporary-salient relationships and patterns, whereas older populations will not, and on those specific sources of s, will appear to perform very poorly relative to

Michael A. Woodley and Aurelio José Figueredo

the average of the contemporary population, even if they have comparable or higher levels of performance on information processing speed measures.

It is important to note that we used these estimates as relative proportions of the Flynn Effect (relative fractions of contemporary estimates for *s.e*) across historical time periods, but did not interpret them as being scaled in the metric of raw IQ scores. Hence an estimated Crepin and Oesterdiekhoff Flynn Effect of 98 becomes 0.98, or 98% of the contemporary value for *s.e* (scaled arbitrarily at 1.00), and this same metric was applied to historical and contemporary estimates for *g.h*. This method avoids a significant problem that arises from comparing between historical *g.h* and *s.e* estimates in terms of raw IQ scores, where the numbers are substantively out of alignment. As proportions of the contemporary mean, these estimates can instead be conceptualized in terms of the relative differences between contemporary and historical populations in the separate domains of *g.h* and *s.e*. The normed Crepin and Oesterdiekhoff baseline of 0.50 was therefore used in estimating the 1455 *s.e* level. With a three point a decade secular increase in the 20th century, the 1900 Western *s.e* would have been about 0.70 (Flynn, 2009a; Oesterdiekhoff, 2012). For both the *g.h* and *s.e* estimates, a contemporary standard of 1.00 was assumed, as normed for 2005.

Eminent individuals in the history of science and technology come from a list of 4,002 names compiled by Murray (2003), who following Galton's (1869) method of 'convergent bibliography', computes this variable on the basis of comparing name prominence across multiple, independently compiled lists of significant figures (encyclopaedias, reference works etc.). Murray then weights this based on both total and 'de facto' (i.e. the proportion of those with access to education, wealth, the larger culture and urban areas – all of which are necessary for enabling innovation) population size so as to derive realistic per capita estimates.

Significant innovation rates are obtained from Huebner (2005), who in turn derived his estimates from a list of 7,198 important events from a comprehensive encyclopaedia of science and technology (Bunch & Hellemans, 2004) that were also weighted to world population size. These estimates exhibit high convergent validity, as they correlate significantly in time >0.80 with alternative estimates compiled by both Murray (2003) and Gary (1993; Woodley, 2012a).

All these data are available for an N of 45 decades (between 1455 and 1945), which is a smaller number of decades than was used in Woodley (2012a; 45 vs. 55 decades). This functions as a statistical control on two fronts. Firstly, Murray (2003) has noted that the proliferation of media post 1950 has distorted people's perceptions of eminence by giving undue relative prominence to comparatively less accomplished individuals than would have been the case in the past. Huebner (2005) has also observed that innovation indices suffer from the inclusion of excessive numbers of less significant or 'micro' innovations in recent decades, which has the effect of attenuating recent declines in innovation rates. Including a more restrictive time series in the analysis substantively reduces the effects of these two potential sources of 'epochcentric inflation'. Secondly, a smaller N reduces the likelihood of Type 1 errors and also the effects of temporal auto-correlation. All data are available in Appendix A.

2.2 Statistical Analysis

All analyses were conducted in SAS 9.3 (2011). We used a combination of multilevel modeling (MLM) for the growth curve analyses and structural equations modeling (SEM) for the path analyses. MLM analyses were performed to formally test for the presence or absence of confounding serially autoregressive effects (i.e. the non-independence of data points in time), and the SEM analyses were used on the unadjusted data when no evidence was found for such effects. This two-pronged approach avoided the estimation, otherwise necessary (Cook & Campbell, 1974), of all the serially autoregressive effects of successive measures of the same variable on itself, which would have necessitated the estimation of many more model parameters than would have been supported by the present sample size. This combination of methods thus enabled greater statistical power for testing the effect of how the changes over time represented by one variable influenced the changes represented by another.

Path analysis, or manifest variable structural equations modeling, consists of imposing a restricted set of causal pathways, also specified *a priori*, and testing the covariances reproduced by these hypothesized causal influences against the covariances observed between the variables. A saturated structural model is merely one that freely estimates the direct correlations between all of the variables; any structural model that can adequately reproduce that pattern of observed correlations with a reduced set of hypothesized causal pathways is deemed to be superior by the principle of parsimony. Structural equations modeling permits the modeling of observed covariances by any combination of direct effects, indirect effects, spurious effects, and residual effects (James, Mulaik & Brett, 1982).

SEM, however, goes beyond testing the adequacy of each model separately and permits one to compare the relative adequacy of two or more models that are "hierarchically nested" within each other, meaning that the models are equivalent but for the more restrictive constraints imposed on one with respect to the other. The more constrained of the models is referred to as the 'restricted' model and the less constrained is referred to as the 'inclusive model'. Several kinds of model constraints are possible. The simplest is to constrain one or more of the path coefficients to zero, which essentially eliminates the pathways so constrained from the model. An alternative tactic for increasing model parsimony is to constrain two or more of the path coefficients to be equal to each other (rather than to zero), thus modeling a hypothesized scenario in which the magnitudes of the causal influences that they represent are estimated as equivalent and thereby reducing the total number of parameter estimates needed in the model. Both of these types of constraint were tested in the present analyses. There is another type of model constraint called a cross-sample equality constraint that is used to test for the statistical equivalence of path coefficients across independent samples, but this type was not relevant to the present data because we were only modeling a single sample.

In such cases the restricted (more constrained) model is compared to the inclusive (less constrained) model by the statistical and practical indices of 'goodness-of-fit' which in turn constitutes the principle criterion of adequacy against which the models are being evaluated. The statistical index of fit used to evaluate our path models was chi-squared, which indicates whether a given model perfectly predicts all of the observed covariances to within the margin for sampling error expected by statistical theory. The practical indices of fit used were the Bentler-Bonett Normed Fit Index (NFI) and the Comparative Fit Index (CFI), for which values greater than 0.90 are generally considered satisfactory levels of practical goodness-of-fit, although some prefer a higher criterion of 0.95 (Bentler & Bonnett, 1980; Bentler, 1995). Of these practical fit indices, the CFIs were given greater weight in evaluation of model adequacy than the NFIs because they have the added virtue of being adjusted for model parsimony and have also been shown to perform well with moderate to small sample sizes (N < 250), especially with Maximum Likelihood estimation (Bentler, 1990; Hu & Bentler, 1995). Alternative fit indices, such as the Bentler-Bonett Non-Normed Fit Index (NNFI), provide poor estimates of model fit with smaller samples (Hu & Bentler, 1999). The differences between hierarchically nested models in their statistical and practical indices of fit indicate the relative loss of fit of the model to the data entailed by either the elimination or constraining of specific causal pathways.

For the hierarchically nested model comparisons, one merely tests the difference in the chi-squared values of the two models against the difference in residual degrees of freedom produced by having imposed the constraint. This permits one to perform tests of the statistical significance of the difference between the models in adequacy of fit because differences between chi-squared values are themselves distributed as chi-squared in that their probability distributions under the null hypothesis are mathematically identical. Differences in NFI and CFI are typically evaluated by a practical, rather than statistical rule of thumb, in which one will tolerate an increase of +0.01 for every residual degree of freedom gained by the constraints being evaluated. Non-significant effects are therefore eliminated from the final, or "restricted", causal models both to achieve greater model parsimony and to enhance the efficiency of parameter estimation. However, these constraints should be designed to test *a priori* hypotheses, and no *post hoc* elimination of causal pathways or fitting of additional model parameters is usually recommended. No such atheoretical model respecifications were utilized in the present analyses.

3. RESULTS

3.1 Univariate Analyses: Growth Curve Models

All univariate growth curve analyses were conducted using SAS PROC MIXED and PROC GLM; MLMs were used to investigate the possibility of serial dependencies (autoregressive effects) among successive time points, using Maximum Likelihood Estimation (MLE). As we lacked sufficient *a priori* theory to specify with any certainty a single covariance structure among the potentially correlated errors, we examined three different common assumptions: (UN) Unstructured, (CS) Compound Symmetry, and (AR1) Autoregression with a Lag of 1. Finally, we compared these results to a simple General Linear Model using Ordinary Least Squares (OLS) estimation. As we expected all four time trends to be curvilinear, we modeled all four major variables of interest (Heritable General Intelligence, Environmentally-Influenced Specialized Mental Abilities, Eminent Individuals, and Innovation Rates) as quadratic in function form, by entering Time (T) and Time-Squared (T^2), as measured in decades, hierarchically as predictors of each criterion.

All four methods of estimation yielded identical parameter estimates to three decimal places. This was because all three covariance structures tested estimated correlations of zero among the residuals, rendering all three MLE methods of estimation essentially equivalent to OLS. This indicated that the four growth curves (one for each of our variables) were significantly different from each other but that there were no serial autoregressive effects on the scale of decades. The four quadratic growth curves were obtained as separate GLM OLS estimates for simplicity (given the

41

mutual discriminability of the growth curve parameters among the variables and the lack of serial dependencies in the data), these results were tabulated in Table 3 and are presented graphically in Figure 3 as standardized distributions rather than as growth curves.

Table 3. Ordinary Least Squares (OLS) Estimates of Linear and Quadratic Time Trends for Hierarchical General Linear Models (GLM/SS1) of Heritable General Intelligence (*g.h*), Environmentally-Influenced Specialized Abilities (*s.e*), Rates of Eminent Individuals, and Rates of Significant Innovations.

	Estimate	SE	F(2,48)	P(Ho)
Heritable General Intelligence	R^2=0.974		909.10	0.0001
T	5.859	0.611	1753.96	0.0001
T^2	-4.893	0.611	-64.23	0.0001
Environmentally-Influenced Specialized Abilities	R^2=0.893		201.10	0.0001
T	-5.547	1.243	375.16	0.0001
T^2	6.465	1.243	27.04	0.0001
Eminence	R^2=0.817		107.43	0.0001
T	7.390	1.627	198.77	0.0001
T^2	-6.525	1.627	16.09	0.0002
Rates of Innovation	R^2=0.723		62.61	0.0001
T	3.143	2.004	123.90	0.0001
T^2	-2.299	2.004	1.32	0.2570

To maximize comparability in Figure 3, the predicted scores plotted for all criterion variables were standardized. The Environmentally-Influenced Specialized Abilities (*s.e*) variable was found to be monotonically increasing and positively accelerated, whereas Heritable General Intelligence (*g.h*), Rates of Eminence, and Rates of Innovation were found to be monotonically increasing and negatively accelerated.

In spite of the finding that the quadratic term for Rates of Innovation was not statistically significant in this sample, we left all variables in quadratic form for the purpose of direct comparison. Contrary to some widespread misconceptions, parametric statistics does not assume that the raw data need be normally distributed: the normality assumption applies exclusively to the residuals. Otherwise, methods such as ANOVA would automatically violate normality by employing categorical (non-normal) predictors. For the residuals to be normally distributed, it is only necessary that the distributions of the predictor(s) and criterion be of the same form (*isomorphic*), meaning that if the former is curvilinear then the latter should be similarly curvilinear and subtracting the weighted predictor(s) from the criterion score will leave nothing but a symmetrical pattern of random errors, centered around zero. This makes sense if one variable, which happens to be curvilinear, is in fact causally influencing the other. Any mismatch between the distribution forms of the criterion with respect to the predictor(s) therefore presents more of a potential problem than that of either variable. The high squared multiple correlations (R^2) reported in Table 3 indicate that the quadratic transformations generally did a very good job in modeling the analytic forms of the temporal functions of all four criterion variables.

3.2 Multivariate Analyses: Structural Equation Models

Structural equations models were constructed with the purpose of using estimated variance components based on the estimates of *g.h* and *s.e* to predict both Rates of Eminence and Rates of Innovation over the specified period of historical time, ranging over the five centuries from 1450-1950.

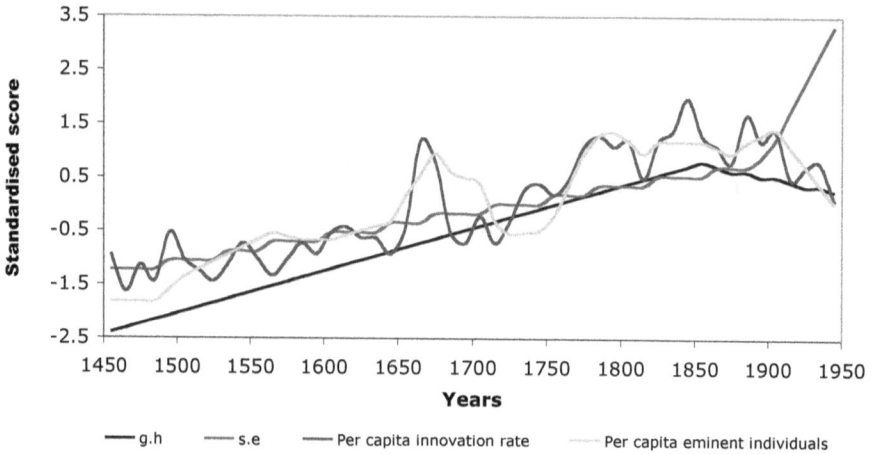

Figure 3. Standardized Distributions for Heritable General Intelligence (*g.h*), Environmentally-Influenced Specialized Mental Abilities (*s.e*), Rates of Eminent Individuals, and Rates of Significant Innovations.

Table 4 presents the results of the nested model comparisons among the alternative structural equation models designed to test the plausibility of varying *a priori* hypotheses about the causal relations among the variables. In the 'inclusive model', both *g.h* and *s.e* were used to separately predict both eminence and innovation rates. The inclusive model, however, was also *saturated*, in that it contained a pathway between every variable and every other, and was therefore not statistically testable because a saturated path-analytic model will by definition always perfectly reproduce the data. Nevertheless, the individual model parameters could be tested and it turned out that both of the specified pathways from *s.e* (to Eminence and Innovation) were negative in direction, negligible in magnitude, and statistically non-significant. We were not aware of this particular result, however, when we had planned out the nested model comparisons *a priori*, so the full set of planned model comparisons will be reported for the sake of completeness and of intellectual integrity.

Two series of restricted models were constructed, estimated, tested, and compared. The alternative model specifications of both series of restricted models are displayed in Figure 4, in relation to the inclusive model (bottom right).

The first series (denoted Series A), including SEM.A1 – SEM.A3, imposed equality constraints upon the corresponding effects of *g.h* and *s.e* on Eminence and Innovation. SEM.A1 was the most restricted of this series, in that such equality constraints were imposed upon the corresponding pathways of *g.h* and *s.e* to both *Eminence* and *Innovation*; SEM.A2 was somewhat less restricted, in that the equality constraint was imposed only upon the corresponding pathways from *g.h* and *s.e* to *Eminence*, but the corresponding pathways from *g.h* and *s.e* to *Innovation* were freely estimated as potentially unequal to each other; SEM.A3 was also less restricted that SEM.A1, but in the opposite sense than the equality constraint was imposed only upon the corresponding pathways from *g.h* and *s.e* to *Innovation*, but the corresponding pathways from *g.h* and *s.e* to *Eminence* were freely estimated as potentially unequal to each other.

The second series (denoted Series B), including SEM.B1 – SEM.B2, instead simply eliminated (constrained to zero) the corresponding effects of *s.e* on both Eminence and Innovation. SEM.B1 was the less restricted of the two, in that no equality constraints were imposed upon the corresponding pathways from *g.h* to *Eminence* and from *g.h* to *Innovation*, and these causal pathway coefficients were freely estimated as potentially unequal; SEM.B2 was more restricted in that an equality constraint was imposed on those two pathways, restricting the parameter estimate for the causal pathway coefficients from *g.h* to *Eminence* and the one from *g.h* to *Innovation* to be estimated as statistically equivalent by means of a single model parameter.

The absolute and relative fit indices of these five models are displayed in Table 4. All models but SEM.B1 were statistically rejectable by the chi-squared criterion, although SEM.A3 and SEM.B2 were acceptable in absolute terms by both practical indices of fit (NFI and CFI). The others were rejectable by all criteria evaluated. Recall, however, that the differences between hierarchically nested models in their statistical and practical indices of fit indicate the relative loss of fit of the model to the data entailed by either the elimination or constraining of specific causal pathways. Within Series A, the nested model comparisons revealed that imposing an additional equality constraint upon the corresponding pathways of *g.h* and *s.e* to either endogenous variable (whether *Eminence* or *Innovation*) after having imposed that same constraint upon the other was rejectable by all statistical and practical criteria evaluated regardless of which endogenous variable was so constrained initially. Within Series B, the nested model comparison revealed that imposing an equality constraint upon the corresponding pathways from *g.h* to *Eminence* and from *g.h* to *Innovation* was also rejectable by all statistical and practical criteria evaluated.

The last and only model left standing from among the tested alternative specifications was therefore SEM.B1, in which causal pathways from *g.h* to *Eminence* and from *g.h* to *Innovation* were left free and unequal to each other, and the corresponding causal pathways from *s.e* to *Eminence* and from *s.e* to *Innovation* were constrained to be zero, meaning entirely eliminated from the model.

The elimination of these two causal pathways from *s.e* to the endogenous variables had no effect on the remaining structural relations modeled, nor did it substantively alter the model fit to the data. Nevertheless, SEM.B1 increased the residual degrees of freedom by two, indicating a gain in model parsimony without

Figure 4. Alternative Structural Equation Models for Heritable General Intelligence (*g.h*), Environmentally-Influenced Specialized Abilities (*s.e*), Rates of Eminent Individuals, and Rates of Significant Innovations.

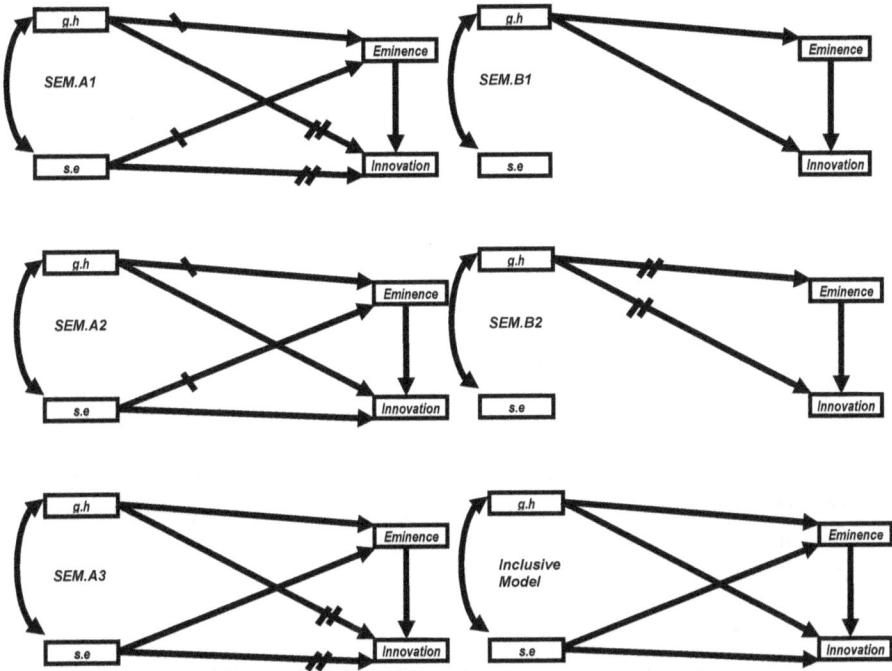

significant loss of explanatory power. On that basis, SEM.B1 was therefore tentatively accepted as the preferred model among the alternative specifications tested, and is presented with Maximum Likelihood Estimation (MLE) parameter estimates in Figure 5. An asterisk (*) indicates that the corresponding model parameter is different from zero at a statistical significance level of $p<0.05$.

Table 4. Hierarchically Nested Structural Equation Models for Heritable General Intelligence (*g.h*), Environmentally-Influenced Specialized Abilities (*s.e*), Rates of Eminent Individuals, and Rates of Significant Innovations *$p<0.05$.

Structural Model	χ^2	RDF	p(Ho)	NFI	CFI
SEM.A1	38.194*	2	0.0001	0.830	0.834
SEM.A2	33.031*	1	0.0001	0.853	0.853
SEM.A3	5.163*	1	0.0231	0.977	0.981
SEM.B1	3.289^{ns}	2	0.1932	0.985	0.994
SEM.B2	9.563*	3	0.0227	0.957	0.970
(SEM.A1 - SEM.A2)	5.163*	1	0.0231	-0.023	-0.019
(SEM.A1 - SEM.A3)	33.031*	1	0.0001	-0.147	-0.147
(SEM.B2 - SEM.B1)	6.274*	1	0.0123	-0.028	-0.024

Figure 5. Structural Equation Model for Causal Relations among Heritable General Intelligence (*g.h*), Environmentally-Influenced Specialized Mental Abilities (*s.e*), Rates of Eminent Individuals, and Rates of Significant Innovations.

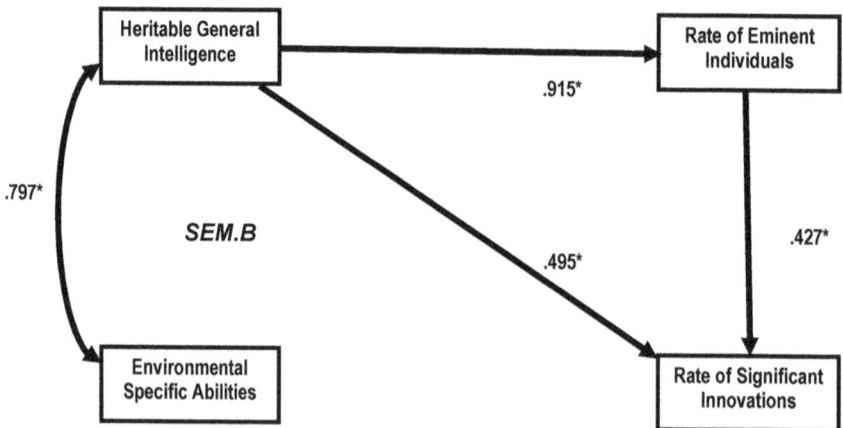

Finally, a feature common to all of these models is the significant causal pathway estimated from *Eminence* to *Innovation*, indicating that a relatively small proportion of "eminent" individuals do indeed contribute disproportionately to the societal rates of significant innovation.

4. DISCUSSION

These findings are quite consistent with the smart fraction theory as they indicate that the effects of mean changes in heritable general intelligence ($g.h$) on per capita innovation rates are partly mediated by the proportion of eminent individuals in the population, thus extending cross-sectionally, similar findings at the cross-national level (Rindermann, Sailer & Thompson, 2009; Rindermann & Thompson, 2011b). Furthermore our measure of the secular increase amongst specialized, high-environmentality abilities ($s.e$ the Flynn effect) does not predict either eminence or significant innovation rates, both of which have been trending in an opposing direction to the Flynn effect since the mid-late 19th century. This is consistent with the observation that Flynn effects, unlike dysgenesic fertility, do not occur for the most part on g (te Nijenhuis & van der Flier, in press; Wicherts et al., 2004).

Complimentary to this is the finding that what evidence there is for the Flynn effect having had real world consequences revolves around observations of increased cognitive specialization, such as precociousness in games like chess, bridge and go or teacher ratings, which indicate that student performance in the domain of 'practical intelligence' has been increasing, rather than overall performance (Howard, 1999, 2001). As the capacity for populations to specialize cognitively over time is apparently unrelated to levels of $g.h$ (it might instead be more dependent upon changes in conative [personality, psycho-social and behavioral traits] rather than cognitive factors: Ackerman, 1996; Ackerman & Beier, 2003; Woodley, 2011a, 2012a, b; Woodley, Figueredo, Brown & Ross in press), both effects can occur simultaneously consistent with the co-occurrence model. The end result of this process is social improvement along performance-dimensions that are contingent upon increasing specialization (e.g. exploiting Ricardo's Law of Comparative Advantage as a means of generating wealth, perhaps via the proliferation of 'micro-level' innovations), coupled with social 'regression' along performance-dimensions that are contingent upon cognitive 'heavy lifting' (such as significant innovation; Woodley, 2012a). It needs to be noted that this interpretation is also consistent with the economic historical research of Robert Fogel (1964), who found that economic growth in the US during the 19th century, was principally determined by the aggregate of large numbers of specific technological changes (what might be considered analogous to 'micro-level' innovations), rather than by the presence of relatively fewer significant (or macro) innovations.

The finding of this and previous studies therefore effectively falsifies claims made by proponents of the attenuation model, or the idea that the Flynn effect is offsetting any socially negative influences from dysgenesis (Bostrom, 2002; Burt, 1948; Loehlin, 1997; Lynn, 1996, 2011), as the capacity of a society for significant innovation appears to be *completely* independent of the Flynn effect. Woodley (2012a) previously found it to be a positive predictor of small magnitude, however one possibility is that the high collinearity between the Flynn effect and the temporal correlation that had been statistically controlled had functioned to obviate this relation here, suggesting that it may have been artificial.

These results also substantively expand on the core finding of Woodley's (2012a) study, as they indicate that the relation of $g.h$ to innovation rates is robust to the effects of controlling for serial autocorrelation, and that furthermore it withstands the inclusion of novel model parameters such as reduced N, the inclusion of a novel variable (eminent individuals), and the use of a more robust estimate of dysgenesis rates.

4.1 Limitations of the Study

The principal limitation of this analysis is that although our two exogenous variables, Rate of Eminent Individuals, and Rates of Significant Innovations, were directly measured, the estimated variance components $g.h$ and $s.e$ could only be inferred from historical sources. Furthermore, our interpretations of these historical trends are based on a specific theoretical framework, which we used as a heuristic in these various reconstructions.

As was discussed in the introduction, it has been argued that, at least in the UK, the rate of dysgenesis was greater in the decades of the late 19th and early 20th centuries (during the demographic transition), than in later decades. Lynn's (1996, 2011) estimates of these differentials can be seen in Table 2. The possibility therefore exists that at the very fine temporal scale dysgenesis rates may be much more stochastic, with some countries experiencing especially pronounced periods of dysgenesis or eugenesis at various points in time owing to the operation of different factors. These might include changes with time in the size of the contraceptive usage differential between the various cognitive classes (Lynn, 1996, 2011), and other differences associated with the fact that not all Western countries went through the demographic transition at precisely the same time (Caldwell, 2001). Another significant factor is associated with the influence of totalitarian regimes, whose policies may have had dysgenic effects (such as in the case of the Nazi Holocaust and political mass murders in Revolutionary France and former Communist Block countries, all of which explicitly targeted for destruction groups exhibiting much higher IQ-means than the population; Glad, 1998; Graham 1998; Itzkoff, 2009; Sternberg, 2005) amongst other factors. As small fluctuations in IQ can have profound effects on the numbers of eminent individuals, it is possible that the stochasticity of these numbers represents an underlying finer-grained stochasticity in $g.h$. Another issue concerns the accuracy of the decline estimates employed here. These are based on the use of pen-and-paper tests that are sensitive to the Flynn effect. Given that dysgenic fertility is basically a monotonic Jensen effect, this suggests that the apparent magnitude of dysgenic fertility will be extremely sensitive to decreases in the g-saturation of the tests with respect to which it is measured. If the Flynn effect is reducing the g-loadings of tests over time, this could indicate that the apparent secular decline in the strength of dysgenic fertility described by Lynn (1996, 2011) is little more than an artefact of the Flynn effect. *The real dysgenesis rate on g may therefore be much higher than the estimates employed here suggest.* An additional confound concerns the possibility that decline estimates inferred from the negative correlation between IQ and fertility are sensitive to only one source of dysgenesis, i.e. the change in the frequencies of genes for g entailed by negative directional selection. It has also been argued that the frequencies of low-g genes are increasing independently owing to the accumulation of pleiotropic mutations with large phenotypic effects entailed by the unprecedented extent to which selection has been relaxed for childhood and early adult diseases, especially in the post-industrial revolution world (Charlton, 2013; Crabtree, 2012a, b; cf. Hamilton, 2002). This combination of reverse individual-level selection for low-g coupled with relaxed selection against pleiotropic mutations with potentially large effects on g therefore constitutes yet another criterion by which the actual dysgenic effect on g might be substantively larger than is revealed in studies investigating the IQ to fertility relationship (Charlton, 2013). Consistent with this, and based on Silverman's (2010) simple reaction time means data, Charlton (2012a) has speculated that g may have declined by as much as a standard deviation between the 1880's and the present day. To properly test this speculation it would be necessary to convert the increase in simple reaction time latency into an equivalent g-decline employing a psychometric meta-analytic procedure in which artefacts such as range restriction and imperfect construct measurement could be properly controlled (Hunter & Schmidt, 2004). Elementary cognitive tasks, show no affinity for the Flynn effect (Nettelbeck & C. Wilson, 2004; Silverman, 2010) hence, unlike the kind of items commonly found on written tests, they are measurement invariant with respect to g over time. Such an analysis is potentially extremely important, as secular trends with respect to these measures are capable of revealing true secular trends in g.

Furthermore, whilst the results of Hart's (2007) computer simulations involving different world regions provide somewhat plausible proxy estimates for historical levels of $g.h$, a better method will eventually be required in order to estimate the levels of this variable in historical time. Perhaps this can be achieved via the examination of historical trends in heritable endophenotypic proxies for $g.h$, such as endocranial volume (Rushton & Ankney, 2009), or via measures of factors such as skills premiums, which capture the degree to which high-g professionals were in demand, hence might serve as a proxy for the historical scarcity/abundance of such individuals. Another possibility is that the data on the temporal variation in the fertility to status relation derived by Skirbekk (2008) could be used to infer historical levels of g based on the well-established contemporary relations between this variable and factors like socio-economic status and education. Future research should therefore aim to construct better estimates of historical levels of $g.h$ taking all of these factors into consideration.

4.2 Objections: Alternative Models of Declining Rates of Innovation

Interestingly, there appears to be something of an 'emerging consensus' on the reality of declining per capita levels of significant innovation (e.g. Arbesman, 2011; Cowen, 2011; Horgan, 1997; Huebner, 2005; Jones, 2009; Kasparov, Levchin & Thiel, 2012; Murray, 2003). However most researchers in this field attribute the declines to causes other than dysgenesis. These alternatives include the 'low hanging fruit' model, which suggests that innovation is getting harder simply because we are exhausting the space of possible innovation (Arbesman, 2011). Another model is that 'economic limits' are starting to exert their toll in as much as the profitability of further innovation is subject to diminishing returns (e.g. Cowen, 2011; Huebner, 2005). Variations on this theme include the idea that growth in the state sector has 'crowded out' private sector funding for the sciences, which has in turn negatively impacted innovation (Kasparov, Levchin & Thiel, 2012; Kealey, 1997), or alternatively that some combination of overly strict or poorly enforced intellectual property rights have stifled innovation in technology (e.g. Panagopoulos, 2004; Qian, 2007).

Yet another model posits that it is in fact 'cultural malaise' that is behind the decline in innovation, as people are increasingly subordinating rigorous personal standards of scientific enquiry to other ends (such as their careers, ideologies or bureaucratic norms; Charlton, 2012b; Murray, 2003). The problem with these alternatives is that they often assume that the intrinsic quality of human capital over time is either constant or increasing, and that the causes of the declines must therefore be attributable to extrinsic factors. We will here argue that many of these extrinsic factor driven models are nonetheless compatible with the reality of a dysgenic trend. A low hanging fruit may for example be within easy reach of someone of sufficient intellectual 'stature', but not to those of lower 'stature'. A decline in intellectual 'stature' could easily therefore be the root cause of once low hanging fruit no longer being within grasping distance.

It is not likely therefore that the space of major scientific and technological innovation is simply becoming generally exhausted, as Horgan (1997) and Arbesman (2011) have argued. In some sub-fields it is true that the innovation rate has slowed or even ceased simply because there are fewer or no discoveries left to be made (such as new stable chemical elements or large species of animal; Arbesman, 2011; Paxton, 1998), however these are a minority of cases. In even fewer instances does it appear to be the case that physical limits on technology have been reached. An example might include the observation that interest in the field of perpetual motion research ground to a halt for the most part with the realization that the laws of thermodynamics made such endeavours impossible. However, apparent physical limits are not always real physical limits. The physical limits imposed on computer processing efficiency by the use of triodes for example did not prevent William Shockley from developing the transistor, and in so doing altering the physical limits on processing massively. The slowing of major scientific progress has affected fields as broad as medicine and cosmology (Charlton, 2012b; Healey, 2013 Simonton, 2013) where the potential for significant novel innovation is undoubtedly still great (Maddox, 1999). The key issue therefore appears to be the deficit of genius coupled with a culture of science in which non-truth seeking orientations prevail (more will be said about this issue subsequently) rather than the absence of potential novelty (Charlton, 2012b; Healey, 2013; Murray, 2003).

Similarly the location of economic 'limits' on technology might owe much to the intellectual capabilities of the populace. An intelligent population is needed not only to produce a smart fraction capable of generating the innovation but also to *effectively utilize* it, and in so doing make its dissemination economically feasible. Evidence for this comes from the presence of a significant direct path from $g.h$ to innovation rates in our model. Furthermore, there is no reason to believe that the application of intelligence to making innovation more cost effective does not yield results – the history of the automobile, the airplane and the computer being pertinent examples of technologies that started out prohibitively expensive, highly inefficient and available to only a handful of individuals, but which now enjoy broad based and efficient usage, all because of certain significant innovations (i.e. the development of petroleum fractionation, the development of the jet engine, the development of the transistor, the personal computer etc).

Perhaps the space industry serves as a good prospective example of dysgenesis increasing the severity of the economic limits imposed on a technology. The argument has been made that the space industry is currently in disarray primarily due to the absence of nationalistic tensions between superpowers jostling for world supremacy providing the necessary incentives (Albrecht, 2011). It is also the case that astronautics and space engineering are likely to be more sensitive to changes in smart fraction size than most innovation sectors, owing principally to the highly g-loaded nature of the challenges confronted by people working in these sectors. Absent the maintenance of a 'critical' smart fraction size the space industry may simply therefore have collapsed. It is remarkable for example that the most efficient space drive ever developed, the nuclear pulse engine, which relies for its propulsive power on the simple principal of detonating a chain of nuclear explosives behind a sturdy enough and well shielded vehicle, is a product of a handful of brilliant engineers and physicists from the 1950s, including Stanislaw Ulam and Freeman Dyson (Dyson,

2003). It is doubly remarkable that this simple and elegant space-drive has never been constructed, despite an apparent desire for manned exploration of the solar system. Furthermore, it has been noted by the US National Academy of Sciences that the mean age of the NASA workforce is increasing as the organization recruits less and less from amongst younger demographics (Committee on Meeting the Workforce Needs for the National Vision for Space Exploration, National Research Council, 2007). It could be that NASA simply isn't attractive enough or doesn't have the resources to recruit the best and the brightest from among the younger demographic. Another implication of this trend is that the younger generations increasingly lack the necessary talent to be effective in domains where high-g is essential for success. This interpretation is bolstered by the observation of a similar trend amongst the winners of the undisputedly highly g-loaded Nobel Prize, who also appear to be increasing in age over time (Jones & Weinberg, 2011), again implying that younger generations are becoming increasingly less competitive.

An alternative interpretation of the finding that scientific innovation is being deferred until later in life (Jones & Weinberg, 2011) is that the 'burden of knowledge' is now so great that it simply takes longer for individual scientists to specialize sufficiently to be able to exploit novelty (Jones, 2009). Firstly, there is no reason to believe that the 'burden of knowledge' should function as a limiting factor to innovation, as an increased knowledge base should in principal translate into increased opportunities to detect combinatorial novelty and hence innovate (Burke, 2007).

Secondly, whilst the 'burden of knowledge' has undoubtedly been increasing, it is also the case that major scientific breakthroughs are often made by individuals from outside of a particular field (Simonton, 2009). The productivity of such individuals seems to be largely a function of their capacity to perceive novelty, often through the application of more rigorous (i.e. 'harder') and less conventional ways of thinking about problems than those in the host discipline (which is typically 'softer' by comparison) are used to dealing with (Simonton refers to this process as domain-regression). It is this, rather than the acquisition of core disciplinary knowledge (what Simonton refers to as domain-typicality), which predominantly predisposes towards being able to generate scientific novelty. Hence if higher-g domain-regressive types are becoming scarcer, relative to domain-typical types due to dysgenesis, then this could explain why the 'burden of knowledge' is now more of a limiting factor on scientific accomplishment than was the case in the past. The absence of genius means that scientific facts simply accumulate piecemeal rather than become integrated into broad novel theoretical frameworks (Charlton, 2012b).

Whilst plausible, the idea that the 'crowding out' of the private sector by the state sector has inhibited innovation does not entirely accord with the data. The UK did not start to publicly fund science until 1913, whereas the US did not start to do this until 1940 (Kealey, 1997). This is long after the beginning of the decline in both innovation and eminence rates (1850's to 1870's) however. Freer markets were apparently not therefore encouraging innovation during this period of decline. This is not to say that relatively more recent 'crowding out' by state growth hasn't imposed additional opportunity costs on innovation, as has been argued by Terence Kealey (1997), Peter Thiel and others (Kasparov, Levchin & Thiel, 2012), it is simply that the data indicate that the decline in innovation appears to have a *cause* that is independent of this process. We contend that this cause is most likely dysgenesis.

It has been argued that there exists a 'sweet spot' in the level of patent legislation, such that too little intellectual property protection removes economic incentives to innovate by artificially lowering barriers for entry into the market place which in turn makes it too easy for others to simply copy pre-existing innovation. Conversely, too much intellectual property protection also removes the incentive to innovate, as it allows monopolies to form, and for patent legislation to be used in creating artificially high entry barriers in the market place to the detriment those with potentially novel innovations. One possibility therefore is that innovation is highest when societies more closely occupy this 'sweet spot' (Panagopoulos, 2004; Qian, 2007). Firstly it needs to be noted that the concept of intellectual property only applies to a subclass of innovations (i.e. technological), whereas much of the decline is a consequence of diminishing scientific innovation in fields where intellectual property laws do not apply (i.e. pure science). Therefore the 'sweet spot' argument cannot account for the entirety of the decline, but may however contribute to declines specific to the domain of technological innovation (there are well documented historical cases of patent monopolists such as Thomas Edison deliberately inhibiting rival innovation to increase their market share for example; Panagopoulos, 2004). Even though it is a potential co-contributor to declining technological innovation, there exist certain observations, which suggest that its overall contribution is likely small. For example, Clark (2007) notes that during the industrial revolution inventors failed to reap dividends from their inventions *despite* possessing patents, yet the rate of innovation was at an all-time high during the late 18th and early 19th centuries. Clark argues that patents and the growth in intellectual property laws coupled with the concomitant increase in economic incentives to innovate, simply therefore couldn't have been responsible for the up-swell in innovation during this period.

Clark concludes that it must therefore have been an increase in the *supply* of eminent innovators resulting from centuries of differential fitness favoring those with bourgeoisie phenotypes that was driving up the rate of innovation. This is of course compatible with the presence of eugenic fertility trends during this period also, which would have contributed to the supply of innovators via the growth in the smart fraction. The converse of this is that the process of dysgenic fertility therefore entails declines in the rate of technological innovation also.

Another plausible explanation for the decline in rates of innovation is that cultural trends might also be involved. We consider this hypothesis to be a complementary rather than truly alternative one because such cultural trends would be inexorably linked to dysgenesis via both gene-culture co-evolution (Cavalli-Sforza & Feldman, 1981; Feldman & Cavalli-Sforza, 1976; Richerson & Boyd, 2006; Woodley 2006) and negative cultural amplifiers (Meisenberg, 2003), which (as was discussed in the introduction) result from small dysgenic genetic changes within a population leveraging relatively much larger negative cultural changes. Many of the 'dyscultural' tendencies of modernity identified by Murray (2003) and Charlton (2012b) could therefore constitute an extension of dysgenesis into the social and cultural realm. This may be especially true of the apparent abandonment of the 'science as transcendent truth seeking' enterprise that was characteristic of the 18th and 19th centuries, as the recognition of instrumentalism as inimical to science may be dependent not just upon individuals being sufficiently intelligent to realize that this is how things *should* be done, but on the presence of sufficient numbers of like-minded individuals in institutions to *enforce* this as the norm.

Charlton (2008) has argued that scientific institutions underwent a shift in terms of their personnel selection criteria in the mid-20th century, where socially desirable personnel were no longer the brightest, but the most conscientious and hard working. As Conscientiousness and IQ are non-correlated (Ackerman & Heggestad, 1997), selection for one does not entail selection for the other, and may even in some instances select against those with the highest IQs, especially in cases where this is coupled with eccentricity and temperamental personality, as is often the case with scientific geniuses (Charlton, 2004; Eysenck, 1995; Simonton 1999). Another relevant factor has been found by Koriat (2012), whose research reveals that in solving novel problems; individuals in groups typically assign the greatest value to the judgments of those who are confident in their solutions, even when they are wrong. If confident but intellectually mediocre individuals dominate science and technology, then it follows on from this that those who might have novel insights, but who are less conformist and less willing to participate in hype will also become increasingly marginalized from the process of innovation.

This sort of shift permits the continued growth of cultural institutions despite diminishing population IQ. A byproduct of this process is that high-IQ problem solvers are not only no longer required for the success of such institutions, but are actually inimical to their success, hence excluded from the process of innovation. Institutions will then tend to substitute the seeking of transcendent scientific truth with more arbitrarily defined criteria, such as the proliferation of bureaucracy: institutional growth for its own sake. Institutional-level cultural selection favors this state of affairs, as is evidenced by the accelerating growth in scientific bureaucracy (Charlton, 2010), furthermore this institutional selection also potentially encourages scientifically fraudulent behaviors as a form of careerism, which further diminishes the capacity for such institutions to contribute meaningfully to the process of innovation (Charlton, 2012b).

That this is an accurate reflection of the state of affairs, rather than mere hyperbole, has been demonstrated recently in a paper by Fang, Steen and Casadevall (2012), who in surveying 2,047 retracted biomedical and life-science papers found that a) the majority (67.4%) of retractions were due to various forms of scientific misconduct, and b) that the rate of retraction has increased approximately 10-fold since 1975.

At the level of the population as a whole, however, the net effect of negative selection for IQ at the individual level and positive cultural selection for larger institutions dominated by a conscientious-but-dull scientific labor force coupled with a confident-but-dull bureaucratic managerial class, is to diminish competitive efficiency, as science and technology appear to be absolutely central to the economic success of such entities (Clark, 2007; Niosi, 1991).

4.3 Objections: Genetic Arguments

There are two genetic lines of argumentation which are relevant to this discussion, one regarding the invariance in the heritability of general intelligence, and another concerning the argument that the dysgenic effect may have been genetically offset by factors such as regression to the mean, assortative mating, and heterosis.

In response to the concern that the heritability of IQ is not stable over time and should in fact be rising owing to improvements in environmental conditions (e.g. Sundet, Tambs, Magnus & Berg, 2002), we make the assumption that g exhibits a fairly flat *norm of reaction* (i.e. the range of 'acceptable' environments for the full realization of adult genetic potential for g is very large). Evidence for this comes from the observation that heritability estimates for IQ from different studies are remarkably consistent across studies and across time, suggesting no consistent secular trend towards increasing heritability of either the *g.h* or *s.h* component of IQ with improving environmental conditions (Jensen, 1998; Sundet, Tambs, Magnus & Berg, 2002).

This result may be considered somewhat surprising given a) the previously discussed monotonic positive relationship between the heritability of IQ tests and their *g*-loadings, and b) the previously discussed observation that subtests are becoming less *g*-loaded with time owing to the Flynn effect. These two observations suggest that the heritability of IQ should be *decreasing* with time. One implication of the finding of no consistent secular trends with regard to heritability therefore is that the predicted increase in IQ heritability owing to environmental improvements has in fact occurred, and that this has offset the decline in heritability entailed by the secular weakening of *g* over time. Thus the two effects may have cancelled each other out.

In response to the unrealistic proposal that regression to the mean might have offset the effects of dysgenic fertility, we offer the following clarification. The expression 'regression to the mean' has been used, albeit inappropriately, to describe the tendency for the offspring of individuals to become slightly more like the mean of a population in terms of their levels of heritable polygenic traits, such as *g*, with each passing generation. It has been argued that this works for those who are both below and above average in terms of levels of a given trait, and is especially pronounced at the extremes of the normal distribution. Hence it has been proposed that whilst the offspring of those with above average *g* might be regressing towards the societal *g*-mean in addition to becoming less numerous, the relatively more numerous offspring of those with below average *g* will be similarly regressing, albeit upwards. This argument has been used to criticize the idea that dysgenesis can meaningfully occur (e.g. Havender, 1987), however its inadequacies were recognized as long ago as the turn of the 20th century, by

65

Pearson (1903) who demonstrated that with differential fertility for intelligence, the population mean is constantly shifting downwards, hence with every generation regression is to a new and lower mean and is thus dysgenic. It needs to be pointed out that the expression *regression to the mean* has been historically misused with great frequency. Mathematically, regression to the mean is considered to be a statistical artifact produced by measurement error (Upton & Cook, 2006), and not a genetically substantive process occurring in the real world. For example, when one selects individuals with unusually high scores on any given measure, one also inadvertently oversamples scores with positive errors of measurement. If these errors of measurement are normally distributed and have a mean of zero, as is presumed in mathematical statistics, then any subsequent measurement of the same individuals will tend to produce lower scores, and hence retest with scores that are closer to the mean. The opposite happens when one selects individuals with unusually low scores on any given measure, while still producing subsequent scores closer to the population mean upon retesting.

Furthermore in the case of parent-offspring correlations on *g*, oversampling parental scores with positive errors of measurement on IQ, as by selecting those identified as high-*g* individuals based on high observed IQ scores for special study, will produce regression to the mean when assessing the IQ of their offspring, even if the offspring were genetically identical to the parents, given the nature of this statistical artifact. This can be confirmed by retesting the parents themselves, which is rarely done, because one will then no doubt observe regression to the mean of the parental IQ scores in the parents themselves, presumably without having undergone any genetic recombination whatsoever. The proposition that offspring are *necessarily* closer to the mean of the general population in their *actual* latent *g*-factor (as opposed to their observed IQ scores) is therefore a fallacy, especially under conditions of assortative mating.

In response to the alternative proposal that either assortative mating or heterosis might have offset the effects of dysgenic fertility, we offer the following considerations. Assortative mating may widen the gaps between different population strata over time, and in so doing, may increase the variance in IQ, thus offsetting dysgenesis (Havender, 1987). Whilst there is certainly evidence for increasing assortative sociality, or homophily amongst different cognitive classes in the West (e.g. Murray, 2012), a key prediction stemming from the hypothesis that this trend is eugenic in potential is that the distribution in IQ should be increasing over time. A study involving historical IQ distribution data on US populations has failed to corroborate this prediction however (Rowe & Rodgers, 2002). Other studies reporting changes in historical IQ distribution in European countries similarly report no secular increases (e.g. Flynn 1987; Sundet, Barlaug & Torjussen, 2004). Therefore, despite increasing assortative sociality amongst the members of the various cognitive classes, this enhanced homophily has failed to translate into differential survival and reproduction favoring higher IQ and is thus dysgenic. Furthermore, any process increasing the variance in IQ would be rendered irrelevant to the issue of dysgenic fertility if low IQ is simply not being selected against in current environments.

The very converse of assortative mating, heterosis, has been proposed as both a minor and a major cause of the Flynn effect (Jensen, 1998; Mingroni, 2004, 2007), however the tendency for secular gains to be *least* pronounced on the sorts of highly heritable and highly *g*-loaded intelligence measures known to be most sensitive to the effects of both inbreeding depression and heterosis (Nagoshi & R. Johnson, 1986; Rushton, 1999), constitutes a significant challenge to this model (Flynn, 2009a; Woodley, 2011b).

Furthermore hypothetical IQ gains from heterosis are small (around 0.6 points per decade; Mingroni, 2007), whereas Flynn effects have historically been very much larger (around 3 points a decade; Lynn, 2009; Woodley, 2011b). Despite this, heterosis could theoretically have offset some of the dysgenic decline in *g.h*, however studies indicate that the documented levels of assortative mating that Western populations have practiced historically are of insufficient magnitude to produce inbreeding depression, which effectively obviates heterosis as either a cause of the Flynn effect in the West (Flynn, 2009a), or as a potential factor offsetting dysgenesis.

4.4 Evolutionary Considerations

To support the validity of the evolutionary assumption that there were selective pressures in the West favoring eugenic fertility before 1850 and favoring dysgenic fertility afterwards until the present day, it is necessary to show evidence for two critical conditions: (1a) that higher-IQ individuals were more successful in terms of survival and reproduction than lower-IQ individuals from 1450-1850; and (2a) that lower-IQ individuals were more successful in terms of survival and reproduction than higher-IQ individuals from 1850-1950.

In support of Point (1a), and as was discussed previously, it has been amply documented that from the beginning of the early modern era until the first quarter of the 19th Century, the differential fertility gradient in Western Industrial Societies, such as those of the British Isles, had been in favor of high IQ individuals (and/or those with conative and physiological traits associated with what has been termed a 'slow life history strategy', i.e. low time preferences, low expenditure of resources into mating, high expenditure into parenting, health, longevity, etc.; Figueredo,

2009), but reversed itself shortly thereafter (Clark, 2007; Skirbekk, 2008).

In support of Point (2a), we presume that the UK trends identified by Lynn (1996, 2011) are broadly representative of similar trends occurring throughout the West (Nyborg, 2012). This can be defended based on Skirbekk's (2008) analysis of the historical social status and education to fertility relation in Western countries, which indicates that all Western countries had unambiguously transitioned into dysgenic fertility for social status and education (and based on the correlation between these and g – the latter also) by the early to mid 19th century. Prior to this those with the highest social status and education had the largest numbers of surviving offspring (Clark, 2007; Skirbekk, 2008), hence fertility for $g.h$ could have been described as 'eugenic'. Lynn (1996, 2011) argues that by 1850 selection based on high infant mortality had broken down, in addition to which fertility had increased to super-replacement levels amongst those with lower $g.h$. Skirbekk's (2008) data on the social status to fertility relation corroborate the latter, thereby justifying our use of 1850 as the inflection point.

To support the applicability of the multilevel selection model discussed in the introduction to the case of exceptional human intellectual accomplishment, it is necessary to show evidence for two critical evolutionary conditions: (1b) that the "genius fraction" of individuals disproportionately making intellectual contributions to society are either not benefiting personally or are actually sacrificing personal success, and thus putting themselves at a competitive disadvantage in within-group competition between individuals; and (2b) that the societies in which these intellectual products are being generated benefit in comparison with other societies, and thus gain a significant competitive advantage in between-group competition.

In support of point (1b) it has been amply documented by Clark (2007) and others (e.g. Kealey, 1997), that the 'eminent' individuals directly responsible for most of these innovations did not fare well in within-group competition, reaping few rewards for their efforts and often faring materially worse than they had prior to making their intellectual contributions to the group. Simonton (2003) has also reviewed the literature on celibacy and marriage rates amongst geniuses, and has concluded that they fare remarkably poorly in contrast to their less eminent peers. McCurdy (1960) for example found that 55% of historical geniuses never married. Even though this suggests a lack of success in the mating market, failure to marry doesn't necessarily entail failure to reproduce, however Ellis (1926) strikingly found that one out of five British geniuses were celibate. This phenomenon is not restricted to historically recent samples either. Not only did Francis Bacon (1597/1942) in his *Essays and the New Atlantis* observe the phenomenon, but remarkably he intuited the group-selected attributes of the products of creative genius also:

> "He that hath wife and children have given hostages to fortune; for they are impediments to great enterprises, either of virtue or mischief. Certainly the best works, and of *greatest merit for the public*, have proceeded from the unmarried of childless men *which, both in affection and means, have married and endowed the public*" (p. 29, italics added for emphasis).

In support of Point (2b), it has been demonstrated in Clark (2007) and elsewhere (e.g. Niosi, 1992) that high innovation rates confer massive competitive advantages to social groups in relation to other groups, both economically and militarily. This is evidenced by the observation that the period of eugenic fertility-fueled expansion of colonizing European empires across several continents was accelerated by the large volume of technological and social innovations produced by the increasing smart fractions of those populations (Hamilton, 2000). For example, mathematical innovations in navigation and mapmaking were essential to the far-flung naval operations of the Spanish and the Portugese in the expansion of their overseas maritime empires during the early modern era (Crosby, 1997), as they were for the British Royal Navy in subsequent centuries.

In further consideration of Point (2b), we note that the combined population of Europe and America both of which have contributed disproportionately to innovation (Murray, 2003), was around 300 million in 1850, whereas it was around 575 million in 1950 (Cameron, 1993). If we take Gelade's (2008) proposed cutoff point for the smart fraction (IQ ≥ 140), around 1.5% of the population would have been in the smart fraction in 1850, assuming a mean IQ of approximately 108 relative to a 'contemporary' (2005) Western IQ of 100 (based on the dysgenesis rates computed here), whereas 0.5% of the population would have been in the smart fraction in 1950, assuming an IQ mean of approximately 101.5 (also based on the estimates used here). In absolute terms, the numbers of people in the smart fraction shrank: the number was around 4.5 million in 1850, whereas the number was around 2.88 million in 1950. The fact that these populations grew in size substantially during this time interval however indicates that this group-level expansion served to attenuate the full impact of declining IQ on absolute smart fraction size, as evidenced by the observation that the decline is smaller (1.62 million people) than would have been the case had the population of the West remained constant at 300 million (3 million people).

This massive population growth would have had a broader influence on the historical global mean IQ as colonizing populations exhibiting higher mean IQ expanded in number greatly relative to the lower mean IQ populations originally resident in the colonized regions due principally to intergroup conflict and resource predation. This numeric expansion would have carried on for a while even after the onset of dysgenic fertility in the West, as evidenced by the fact that between 1850 and 1900 the population of Africa grew by roughly 50 million people, whereas the population of Europe in contrast grew by nearly 100 million, indicating a 2:1 growth ratio favoring Europeans during this period (Cameron, 1993).

Amongst European powers, from the end of the Napoleonic Wars in 1815 until World War I in 1914, the period sometimes known as the *Pax Britannica* (Pugh, 1999), major wars were reduced to a minimum, although their imperial expansion at the expense of non-Western peoples proceeded apace. The latter source of conflict, however, did not pose a serious threat to the European empires themselves, from the standpoint of intergroup competitive threat, until the empires started subsequently coming into conflict with each other. Thus, the end of what has been described as eugenic fertility and the beginning of what has been described as dysgenic fertility coincides precisely with a change, if not a complete reversal, in the balance between the relative magnitudes of the selective pressures produced by within-group with respect to between-group competition.

A consequence of the cessation of group expansion and the relaxation of intergroup selective pressures was a population increase amongst inhabitants of both historically and contemporaneously lower mean IQ non-Western regions, such as Africa, the Middle-East and South-East Asia (Hart, 2007; Lynn, 2006, 2012; Lynn & Vanhanen, 2012), which resulted in part from investments of infrastructure and scientific medicine into these countries. For example, as was mentioned previously the population of Africa in 1850 was a little under 100 million, whereas the population of Europe was a little over 200 million (Cameron, 1993). The contemporary (c. 2008) population of Africa is 984 million, compared with a European population of 603 million however (International Energy Agency, 2011). This process would have created yet another source of declining mean IQ, only this time in the differential fertility between nations rather than within them (Lynn, 2011; Lynn & Harvey, 2004, Meisenberg, 2009; Reeve, 2009; Shatz, 2008). Researchers have estimated that this process has the potential to reduce the worlds mean IQ by up to -1.34 points per decade (Meisenberg, 2009). Although they do not mention the ramifications for IQ, some researchers have speculated that this process of slowing population growth in developed countries, relative to booming population growth in developing countries might in part be behind the apparent severity of the decline in global innovation rates since the late 19th century (Coates, 2005; Modis, 2005). This process of IQ decline via differential inter-group expansion therefore likely compounded the IQ decline via intra-group dysgenesis.

It also finally needs to be noted that the phenomenon of intra-group dysgenesis is not restricted to ethnically Western countries and peoples, with results reported from Sudan (Lynn & Vanhanen, 2012), Taiwan (H-Y. Chen, Y-H. Chen, Liao & H-P. Chen, in press), Dominica (Meisenberg, Lawless, Lambert & Newton, 2006) and Libya (Abdalgadr Al-Shahomee, Lynn & El-ghmary Abdalla, 2013) indicating dysgenic fertility with respect to IQ in these countries. Meisenberg (2008) has furthermore found negative correlations between education and fertility across a globally representative sample of countries, and Meisenberg and Kaul (2010) found that even within a single country (the US), dysgenesis was present in different ethnic groups, albeit to different degrees. Furthermore Skirbekk (2008) found that the social status-fertility relationship didn't become negative in the developing world until the early decades of the 20th century, however the gradient of the negative relationship seems to be steeper over time in these regions than in the West. This very strongly indicates that dysgenesis is presently global in extent due in some part to the spread of the demographic transition and associated fertility behaviors beyond the West (Caldwell, 2001).

Recall that we are using the words eugenesis and dysgenesis as defined in terms of the historically dynamic balance between individual and group selective pressures over time, and that these descriptions of documented patterns of individual and group behaviors are made here without endorsing or condemning any of them, either explicitly or implicitly.

4.5 The Broader Evolutionary Ecological Context

As a possible source of ultimate ecological causation for these observations, although Clark (2007) only alludes to it in passing, we must also take into consideration that the period from 1450-1850 was largely contiguous with the so-called 'Little Ice Age' in the Northern Hemisphere (Fagan, 2001), which produced disastrous consequences for the European Peasantry and conferred a substantial relative advantage to the 'rural bourgeoisie' that were both cognitively (g) and conatively (i.e. in terms of the complex of behavioral adaptations that constitute a 'slow life history strategy') prepared to adopt the scientific agricultural technologies required to adapt to the altered conditions (Figueredo, 2009). Throughout most of the period from 1450-1850, European agriculture suffered extreme, persistent, and repeated dislocations, producing distress primarily among the peasantry that included periodic famines, widespread hypothermia, bread riots, and increased oppression by despotic leaders.

Fagan (2001) further claims that the massive economic and technological reorganizations that were required for sustained and more efficient food production helped favor the new modes of thought that fueled the Industrial Revolution, which was essentially initiated in the agricultural sector in response to the ecological pressures of climate change. This coincides with Clark's claim that what could be characterized as the bourgeois phenotype was favored during the entire period of the Little Ice Age. After 1850, the hostile conditions afflicting the European agricultural sector were ameliorated by the warming trend that has continued until the present day (Esper et al., 2012), which reversed the direction of this fitness differential.

From an evolutionary perspective, we interpret these effects as influencing the strengths of both within-group and between-group competition and therefore selection. Although a selectionist perspective is one not explicitly taken by Fagan, his descriptions clearly point to heightened *within*-group selection disfavoring the lower-IQ strata of European society in a manner consistent with the fertility differentials among social strata described by Clark. In parallel with these findings, several studies have found significant negative temporal correlations between mean temperatures and various measures of international warfare. For example, the recent studies by Zhang et al. (2007, 2011) provide strong support for heightened *between*-group selection during the same historical period described by Fagan, presumably favoring the more prosocial and altruistic phenotypes, by their elucidation of causal linkages between climate change and successive agricultural, socioeconomic and demographic human crises, culminating in what has been called the General Crisis of the Seventeenth Century. Thus, although Zhang et al. did not take an explicitly selectionist perspective any more than did Fagan, intergroup conflict was unequivocally heightened during this period of aggravated ecological stress, which we interpret as consistent with our model.

Zhang et al. (2007) report temporal correlations ranging from between -0.079 to -0.640 at a variety of different geographical scales (i.e. global to regional) between measures of the number of wars and a temperature reconstruction over the period 1400 to 1900. Furthermore, Zhang et al. (2011) found temporal correlations ranging from -0.397 to -0.602 between two measures of the impact of war and two past temperature reconstructions, over the period 1500 to 1800. A sample of the relations illustrating these effects, extracted from the original graphics and superimposed upon each other, are displayed in Figure 6.

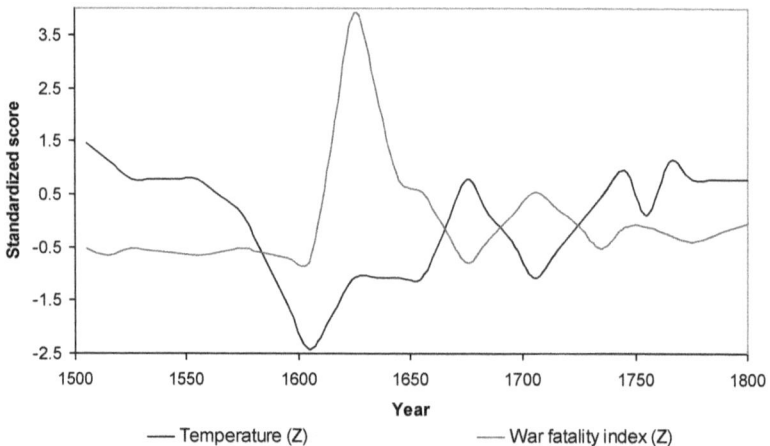

Figure 6. Combined standardized values for European temperature means and war fatality index values taken from sub-figures A and G in figure 1, Zhang et al. (2011, p. 17297). The temporal correlation between these two time series is reported as -0.470 (p<0.0001).

The statistically significant and strongly inverse relation between the war fatalities index and European temperature means of -0.470 reported in Zhang et al. (2011), indicates that this was not a chance association and spanned at least three centuries of what historians refer to as the early modern era.

Taken together, these findings strongly suggest that *natural* selective pressures generated by climate change are what influenced the relative and absolute strengths of the *social* selective pressures in the West described by both Clark and Fagan, although these were depicted in traditional socio-historical terms and not explicitly labeled as such. The warming trend that started in the early 19th Century and continues to the present day therefore ameliorated these two forms of human intraspecific competition and we believe is ultimately responsible for the so-called *dysgenic* effects on *g.h* observed subsequently. This is consistent with the theories of both Lynn (1991, 2006) and Hart (2007) that adaptation to harsher climates was a key selective pressure in the evolution of increased intelligence amongst Western populations. We simply extend that reasoning into historical times.

Finally, it needs to be noted that this process of diminishing harshness, in conjunction with improvements in technology also entailed the relatively more recent runaway environmental and social improvements characteristic of the demographic transition, which further insulated Western populations from sources of environmental and social harshness. In so doing they likely potentiated dysgenic trends. These social buffering mechanisms included generalized education, redistributive economic policy (i.e. welfare states), improvements in both the quality and accessibility of health care, fertility control and family planning. Many of these developments have been directly implicated in dysgenesis (i.e. female education; Meisenberg, 2010; Meisenberg & Kaul, 2010; fertility control, welfareism and enhanced health care; Benson,

2010; Herrnstein & Murray, 1994; Lynn, 1996, 2011; Meisenberg, 2007; Udry, 1978). The enhanced social buffering against harshness afforded by the demographic transition has, consistent with our model, attenuated the potential for inter-group conflict also as evidenced by continuing trends towards diminishing warfare between Western and non-Western populations and also diminishing levels of racism within increasingly multi-racial Western nations. These trends have been accompanied by decreasing societal toleration for war, racism and other forms of inter-group conflict also (Pinker, 2011).

5. CONCLUSIONS

The proposed multilevel selection model implies that 'eminence', or high IQ coupled with sufficiently prosocial motivation to contribute to one's society (perhaps a consequence of slow life history strategy), often with little or no offsetting material compensation, may represent a classic example of self-sacrificing altruism in the evolutionary sense. Further, it suggests that the contributions of 'eminent' individuals might be under positive group-level selection, even if they might simultaneously be under negative individual-level selection. Unusually high intelligence might therefore be a form of group-selected altruism that can only be maintained by sufficiently high levels of between-group competition to offset the fitness costs entailed in the context of within-group competition. Integral to this process was the presence of significant ecological stress such as that resulting from the Little Ice Age, which would have been an important factor in heightening between-group competition and also mortality rates amongst the peasantry.

It finally and importantly needs to be noted that we are advocating neither coercive within-group eugenic measures nor expansionist between-group military policies at the expense of other peoples. Both co-authors of the present monograph publicly self-identify as Libertarians and therefore have extremely strong disinclinations to endorse any such policies. As scientists, however, we are merely describing what might be the fundamental evolutionary dynamics driving the systematic demographic changes that have been widely documented to have occurred throughout the modern era. If our interpretations are correct, this perspective might throw an entirely new light on the phenomena of what have been referred to, perhaps misleadingly, as eugenic and dysgenic fertility.

Acknowledgements

We would like to thank Bruce Charlton and Henry Harpending for their advanced praise of our work, and Henry especially for constructive criticism, for arranging to have the work peer reviewed and for his tireless advocacy. In addition to the two anonymous peer reviewers who evaluated the work at Henry's request, we would like to thank Richard Lynn, Geoffrey Miller, Guy Madison, Gerhard Meisenberg, Richard Gorsuch and James Thompson, who also refereed the work at our request offering up many useful and constructive criticisms along the way. We would also like to thank the faculty and students of the Ethology and Evolutionary Psychology Lab at the University of Arizona for their constructive criticism, and also Elijah L. Armstrong for editorial suggestions. Finally we would like to thank the University of Buckingham Press for agreeing to take on this project.

APPENDIX A

Appendix: A Table of All Data-Sets Used in the Analysis

Decade	Heritable general intelligence (*g.h*)	Flynn effect (s.e)	Rates of significant innovation	Rates of eminent individuals
1455	0.73	0.58	5.5	-1.4
1465	0.74	0.58	2	-1.4
1475	0.75	0.58	4.5	-1.4
1485	0.75	0.58	3	-1.4
1495	0.76	0.59	7.5	-1.2
1505	0.77	0.59	5	-1
1515	0.78	0.59	4	-0.88
1525	0.79	0.59	3	-0.76
1535	0.80	0.60	4.5	-0.64
1545	0.81	0.60	6.5	-0.52
1555	0.82	0.60	5	-0.4
1565	0.82	0.61	3.5	-0.3
1575	0.83	0.61	5	-0.35
1585	0.84	0.61	6.5	-0.4
1595	0.85	0.61	5.5	-0.4
1605	0.86	0.62	7.5	-0.4
1615	0.87	0.62	8	-0.33
1625	0.88	0.62	7	-0.25
1635	0.89	0.62	7	-0.18
1645	0.89	0.63	5.5	-0.1
1655	0.90	0.63	8	0.3
1665	0.91	0.63	16	0.65
1675	0.92	0.64	14	1
1685	0.93	0.64	7.5	0.7
1695	0.94	0.64	6.5	0.6

1705	0.95	0.64	9	0.5
1715	0.96	0.65	6.5	-0.1
1725	0.97	0.65	8.5	-0.3
1735	0.97	0.65	11.5	-0.28
1745	0.98	0.65	12	-0.25
1755	0.99	0.66	11	0
1765	1.00	0.66	12.5	0.5
1775	1.01	0.66	15.5	1
1785	1.02	0.67	16.5	1.3
1795	1.03	0.67	15.5	1.35
1805	1.04	0.67	16	1.2
1815	1.04	0.67	12.5	1
1825	1.05	0.68	16	1.2
1835	1.06	0.68	17	1.2
1845	1.07	0.68	20	1.2
1855	1.08	0.58	16.5	1.2
1865	1.07	0.69	15.5	1.1
1875	1.06	0.69	14	1
1885	1.06	0.69	18.5	1.2
1895	1.05	0.70	16	1.3
1905	1.05	0.72	17	1.4
1915	1.04	0.75	12.5	1.1
1925	1.03	0.78	13.5	0.8
1935	1.03	0.81	14	0.5
1945	1.02	0.84	10.5	0.2

APPENDIX B

Appendix: Apportioning IQ into its Heritable and Environmental Variance Components.

According to the notation used in Table 1, phenotypic aggregate IQ ($IQ.p$) can be partitioned in two complementary ways. First, we may partition $IQ.p$ into components of common factor variance and specific factor variance, as follows:

Equation 1.
$$IQ.p = (IQ.g) + (IQ.s)$$

Equation 2.
$$IQ.g = (p.g)(IQ.p)$$

Equation 3.
$$IQ.s = (1-p.g)(IQ.p)$$

$p.g$ is the proportion of variance in $IQ.p$ that is common factor variance attributable to what has been called *general intelligence*.

Alternatively, we may partition $IQ.p$ by genetically-influenced (heritable) variance and environmentally-influenced variance, as follows:

Equation 4.
$$IQ.p = (IQ.h) + (IQ.e)$$

Equation 5.
$$IQ.h = (p.h)(IQ.p)$$

Equation 6.

$$IQ.e = (1-p.h)(IQ.p)$$

p.h is the proportion of variance in *IQ.p* that is genetically-influenced (heritable) variance attributable to what has been called *heritable intelligence*, which nonetheless continues to include the heritable components of *specific variance* in specialized mental abilities.

In order to estimate the component of intelligence that is both *general* and *heritable* at once (*g.h*) one would have to apply both of the coefficients defined above as follows:

Equation 7.

$$g.h = (p.g)(p.h)(IQ.p)$$

The logical extension of this reasoning is that in order to accurately estimate the aggregate phenotypic IQ from these fundamental components and parameters, one would have to substitute the following composite algorithms:

Equation 8.

$$IQ.p = [g.h] + [g.e] + [s.h] + [s.e]$$

Equation 9.

$$IQ.p = [(p.g)(p.h)(IQ.p)] + [(p.g)(1-p.h)(IQ.p)] + [(1-p.g)(p.h)(IQ.p)] + [(1-p.g)(p.h)(IQ.p)]$$

The final difficulty here, however, is that it is not clear that the same heritability coefficient (*p.h*) applies to both general intelligence and specific mental abilities. There could be two unequal parameters that generate our final heritability coefficient for IQ (*p.h*), which we will denote as *p.h.g* and *p.h.s*, respectively, as follows:

Equation 10.
$$IQ.p = [(p.g)(p.h.g)(IQ.p)] + [(p.g)(1p.h.g)(IQ.p)] + [(1p.g)(p.h.s)(IQ.p)] + [(1-p.g)(p.h.s)(IQ.p)]$$

However, recent papers (W. Johnson et al., 2007; Trzaskowski, Yang, Visscher & Plomin, in press) have estimated the heritability of general intelligence (*g.h*) at between .73 and .77 (utilizing different techniques), which is nearly identical to the heritability of aggregate IQ (*IQ.h*) of ≥.75 (Bouchard Jr., 2004; Gottfredson, 1997; Neisser et al., 1996); this makes it mathematically inevitable that *p.h.g* and *p.h.s* be approximately equal. It is important to note however that this pertains *only* to the aggregate of *p.h.s*. Specific sources of *s*, especially those found in the lowest stratum of Carroll's (1993) hierarchical model of intelligence, will have substantially lower heritabilities than the *p.h.g* component, thus providing the Flynn effect with a substrate on which it can operate in accordance with the co-occurrence model presented here.

Summarizing the major mathematical point that we are making here, however: to correctly estimate the decline in *heritable general intelligence* (*g.h*) from fertility differentials, it is not necessary to multiply the inferred decline by the *heritability* of IQ (*p.h*), as is typically done in estimating what has been called Genotypic IQ (given that only the heritable component of IQ can be plausibly transmitted by differential rates of reproduction and, hence, genetic replication), but instead by the proportion of IQ (*p.g*) that represents the fraction that is *common factor variance* (*g* itself). Although that would be the more precise and theoretically justifiable method of estimation, the currently unknown variability in *p.g* over historical time makes that correction impossible to apply based on the extant psychometric knowledge of the levels that this parameter might have assumed in previous centuries.

GLOSSARY OF NEW TERMS

Attenuation model

A potential solution to 'Cattell's paradox', or the apparent contradiction presented by the presence of secular gains in measured IQ despite the existence of persistent negative IQ-fertility correlations (Higgins, E. Reed & S. Reed, 1962). The model maintains that large secular gains in measured intelligence due to the Flynn effect have masked or *attenuated* comparatively smaller losses in 'Genotypic IQ' due to dysgenesis, both at the level of test-performance *and* in terms of social improvements in factors such as wealth and technological progress. The model has its origins in the work of Sir Cyril Burt (1948) who was the first to propose that environmental improvements might be reversing IQ declines due to dysgenesis. The theory was restated in much more concrete terms by Richard Lynn (1996), who has argued that much of the Flynn effect can be accounted for by improvements in nutrition (Lynn, 1989, 2009). This is taken to imply that the secular gains are themselves occurring at the level of 'Phenotypic IQ', and hence reflect an across the board or general increase in IQ (Jensen, 1998; i.e. both the g and s variance proportions are responding positively to nutritional interventions). The model was succinctly illustrated by John Loehlin (1997), who used the analogy of 'leaky boats' (symbolizing dysgenesis) on 'rising tides' (symbolizing the secular gain in IQ).

Co-occurrence model

An alternative solution to 'Cattell's paradox'. The model maintains that the dysgenic effect on IQ is concentrated on the most g-loaded, most heritable and also the most measurement invariant indicators of cognitive ability (Woodley & Meisenberg, in press). Secular gains are by contrast concentrated on relatively less g-loaded, concomitantly less heritable and also less measurement invariant indicators of cognitive ability (te Nijenhuis & van der Flier, in press; Wicherts et at., 2004). The model was first articulated by Michael Woodley (2012b), and predicts that as dysgenic effects represent an *actual* decline in heritable general intelligence ($g.h$), this in turn has negatively influenced g-dependent social development dimensions, such as per-capita rates of significant innovation (Woodley, 2012a) and the per capita numbers of eminent individuals (as demonstrated here). Conversely secular gains in certain high-environmentality narrow cognitive abilities ($s.e$) have led to improvements with respect to dimensions of social development that are dependent upon the proliferation of cognitive specializations, such as wealth creation via micro-scale innovation and the division of labor (Woodley, 2012a; Woodley, Figueredo, Brown & Ross, in press).

Dysgenic

As reconceptualized in our multi-level selection model, dysgenics (or dysgenesis) refers to the outcome of individual-level selection for traits, such as lower g, that reduce the competitiveness of groups, hence can be said to diminish fitness, albeit at the group, rather than the individual level (Cattell, 1972; Figueredo, 2012). This restatement of dysgenics obviates problems associated with a previous common usage of the term in which it was taken to indicate evolutionarily nonsensical processes such as the 'fittest' being outbred by the 'unfit'. This restatement also permits the term to be stripped of its normative content, as the 'dys' component of the word describes the objectively measurable group level *reduction* in fitness entailed by the process, rather than any kind of value-laden judgements about the 'socially undesirable' proliferation of certain traits.

Dysgenesis syndrome

A nexus that exists amongst variables that have a) been trending with time in a direction consistent with the presence of dysgenesis, and b) all share phenomenology by virtue of the fact that they exhibit the Jensen effect – i.e. their association with IQ is positively mediated by g. These variables include giftedness assessed psychometrically (the decline in which is evidenced by the decline in per-capita eminent individuals, as described in Murray (2003)), simple reaction times (Silverman, 2010), dysgenic fertility itself (Woodley & Meisenberg, in press) and the anti-Flynn effect on pen-and-paper IQ tests, which has recently been found to exhibit some affinity for more g-loaded subtests (Woodley & Meisenberg, 2013). Each component of the syndrome constitutes convergent evidence for the co-occurrence model, as it indicates that g really has been declining *despite* secular gains in measured IQ.

Eugenic

As reconceptualized in our multi-level selection model, eugenics refers to the outcome of individual-level social, natural, sexual or artificial selection for higher g and group-level selection for rare possibly *emergenic* phenotypes, such as genius, which enhance the competitiveness of groups via innovation, *despite* being at an apparent disadvantage in inter-individual competition (Figueredo, 2012; Hamilton, 2000). As with dysgenics, this restatement obviates the problem associated with the contradiction presented by supposedly 'more fit' eugenic phenotypes being manifestly less fit at the level of individual-level selection, and also the problem of subjective value judgements concerning the desirability of certain traits. This is because genuinely eugenic outcomes can be objectively coupled with an observable outcome, i.e. enhanced group competitiveness and concomitant expansion, as occurred in the Age of Empire.

REFERENCES

Abdalgadr Al-Shahomee, A., Lynn, R., & El-ghmary Abdalla, S. (2013). Dysgenic fertility, intelligence and family size in Libya. *Intelligence, 41*, 67-69.

Ackerman, P. (1996). A theory of adult intellectual development: process, personality, interests, and knowledge. *Intelligence, 22,* 227–257.

Ackerman, P., & Beier, M. E. (2003). Trait complexes, cognitive investment and domain knowledge. In R. Sternberg & E. Grigorenko (Eds.), *Perspectives on the psychology of abilities, competencies, and expertise* (pp. 1–30). New York: Cambridge University Press.

Ackerman, P. L., & Heggestad, E. D. (1997). Intelligence, personality, and interests: evidence for overlapping traits. *Psychological Bulletin, 121*, 219–245.

Albrecht, M. (2011). *Falling back to Earth: A first hand account of the great space race and the end of the Cold War.* San Mateo: New Media Books.

Arbesman, S. (2011). Quantifying the ease of scientific discovery. *Scientometrics, 86*, 245–250.

Arneson, J. J., Sackett, P. R., & Beatty, A. S. (2011). Ability-performance relationships in education and employment settings: critical tests of the more-is-better and the good-enough hypotheses. *Psychological Science, 22*, 1336-1342.

Bacon, F. (1597/1942). *Essays and the new Atlantis.* Roslyn: Black

Bartholomew, D. J. (2004). *Measuring intelligence: Facts and fallacies.* Cambridge: Cambridge University Press.

Benson, A. R. (2010). The demographic implications of escalating welfare payments in Australia. *Mankind Quarterly, 51*, 127-138.

Bentler, P. M. (1995). *EQS structural equations program manual.* Encino: Multivariate Software, Inc.

Bentler, P. M. (1990). Comparative fit indexes in structural models. *Psychological Bulletin, 107,* 238-246.

Bentler, P. M., & Bonett, D. G. (1980). Significance tests and goodness of fit in the analysis of covariance structures. *Psychological Bulletin, 88,* 588-606.

Bostrom, N. (2002). Existential risks: analyzing human extinction scenarios and related hazards. *Journal of Evolution and Technology, 9,* http://www.jetpress.org/volume9/risks.html

Bouchard Jr, T. J. (2004). Genetic influence on human psychological traits – a survey. *Current Directions in Psychological Science, 13,* 148-151.

Bradford, E. J. G. (1925). Can present scholastic standards be maintained? *Forum of Education, 3,* 186-194.

Brand, C. R. (1996). *The g factor: General Intelligence and its implications.* Chichester: John Wiley & Sons

Bunch, B., & Hellemans, A. (2004). *The history of science and technology: A browser's guide to the great discoveries, inventions, and the people who made them from the dawn of time to today.* New York: Houghton Mifflin Company.

Burks, B. S., & Jones, H. E. (1935). A study of differential fertility in two Californian cities. *Human Biology, 7,* 539-554.

Burke, J. (2007). *Connections: From Ptolemy's astrolabe to the discovery of electricity: How inventions are linked – and how they cause change throughout history.* New York: Simon & Schuster Paperbacks.

Burt, C. (1948). *Intelligence and fertility: The effect of the differential birthrate on inborn mental characteristics.* London: Eugenics Society.

Caldwell, J. C. (2001). The globalization of fertility behavior. *Population and Development Review, 27* (Supplement: Global Fertility Transition), S249–S271.

Cameron, R. (1993). *Concise economic history of the world.* New York: Oxford University Press.

Carroll, J. B. (1993). *Human cognitive abilities: A survey of factor-analytic studies.* Cambridge: Cambridge University Press.

Cattell, R. B. (1987). *Beyondism: Religion from science.* Westport: Praeger.

Cattell, R. B. (1972). *A new morality from science: Beyondism.* New York: Pergamon.

Cattell, R. B. (1950). The fate of national intelligence: test of a thirteen-year prediction. *The Eugenics Review, 42,* 136–148.

Cattell, R. B. (1937). *The fight for our national intelligence.* London: P. S. King & Son, Ltd.

Cattell, R. B. (1936). Is our national intelligence declining? *The Eugenics Review, 28,* 181-203.

Cavalli-Sforza, L. L., & Feldman, M. (1981). *Cultural transmission and evolution: A quantitative approach.* Princeton: Princeton University Press.

Chapman, J. C., & Wiggins, D. M. (1925). The relation of family size to intelligence of offspring and socio-economic status of family. *Pedagogical Seminary and Journal of Genetic Psychology, 32,* 414-421.

Charlton, B. G. (2013). What are the genetic causes of the 'dysgenic' decline in intelligence over the past couple of centuries? *Intelligence, Personality and Genius.* http://iqpersonalitygenius.blogspot.co.uk/2013/02/what-are-genetic-causes-of-dysgenic.html

Charlton, B. G. (2012a). Objective and direct evidence of 'dysgenic' decline in genetic '*g*'. *Bruce Charlton's Miscellany.* http://charltonteaching.blogspot.co.uk/2012/02/convincing-objective-and-direct.html

Charlton, B. G. (2012b). *Not even trying - the corruption of real science.* Buckingham: University of Buckingham Press.

Charlton, B. G. (2010). The cancer of bureaucracy: how it will destroy science, medicine, education; and eventually everything else. *Medical Hypotheses, 74*, 961-965.

Charlton, B. G. (2008). Why are modern scientists so dull? How science selects for perseverance and sociability at the expense of intelligence and creativity. *Medical Hypotheses, 72*, 237–243.

Charlton, B. G. (2004). The last genius? - reflections on the death of Francis Crick. *Medical Hypotheses, 63*, 923-924.

Chen, H-Y., Chen, Y-H., Liao, Y-K., & Chen, H-P. (In press). Relationship of fertility with intelligence and education in Taiwan: a brief report. *Journal of Biosocial Science.* doi:10.1017/S0021932012000545.

Clark, G. (2007). *A farewell to alms: A brief economic history of the world.* Princeton: Princeton University Press.

Coates, J. (2005). Looking ahead: a visible end to innovation? I think not. *Research Technology Management.* http//:josephcoates.com/pdf_files/282_An_End_to_Innovation.pdf

Committee on Meeting the Workforce Needs for the National Vision for Space Exploration, National Research Council. (2007). *Building a better NASA workforce: Meeting the workforce needs for the national vision for space exploration.* Washington DC: The National Academies Press.

Cook, T. D., & Campbell, D. T. (1974). *Quasi-experimentation: Design and analysis issues for field settings.* Chicago: Rand McNally.

Cotton, S. M., Kiely, P. M., Crewther, D. P., Thomson, B., Laycock, R., & Crewther, S. G. (2005). A normative and reliability study for the Raven's Colored Progressive Matrices for primary school aged children in Australia. *Personality and Individual Differences, 39*, 647–660.

Cowen, T. (2011). *The great stagnation: How America ate all the low-hanging fruit of modern history, got sick, and will (eventually) feel better.* A Penguin eSpecial from Dutton.

Crabtree, G. R. (2012a). Our fragile intellect. Part 1. *Trends in Genetics, 29*, 1-3.

Crabtree, G. R. (2012b). Our fragile intellect. Part 2. *Trends in Genetics, 29*, 3-5.

Crepin, H. (2009). L'effet Flynn: ou une tentative d'analyser l' évolution de l'intelligence au cours de l'Histoire. *ComMensal, 10,* 5–7.

Crosby, A. W. (1997). *The measure of reality: Quantification and Western society, 1250-1600.* Cambridge: Cambridge University Press.

Damrin, D. E. (1949). Family size and sibling age, sex and position as related to certain aspects of adjustment. *Journal of Social Psychology, 29*, 93-102.

Darwin, C. (1871). *The descent of man, and selection in relation to sex.* London: D. Appleton & Co.

Dickens, W. T., & Flynn, J. R. (2001). Heritability estimates versus large environmental effects: the IQ paradox resolved. *Psychological Review, 108,* 346–369.

Dyson, G. (2003). Project Orion – The Atomic Spaceship 1957-1965. New York: Penguin.

Ellis, H. (1926). *A study of British genius* (revised edition). Boston: Houghton Mifflin.

Emanuelsson, I., Reuterberg, S. E., & Svensson, A. (1993). Changing differences in intelligence? Comparisons between groups of 13-year-olds tested from 1960 to 1990. *Scandinavian Journal of Educational Research, 37,* 259–276.

Emmett, W. G. (1950). The trend of intelligence in certain districts of England. *Population Studies, 3,* 324-337.

Eppig, C., Fincher, C. L., & Thornhill, R. (2010). Parasite prevalence and the worldwide distribution of cognitive ability.

Proceedings of the Royal Society, Series B, Biological Sciences, 277, 3801–3808.

Esper, J., Frank, D. C., Timonen, M., Zorita, E., Wilson, R. J. S., Luterbacher, J., Holzkämper, S., Fischer, N., Wagner, S., Nievergelt, D., Verstege, A., & Büntgen, U. (2012). Orbital forcing of tree-ring data. *Nature Climate Change, 2,* 862-866.

Eysenck, H. J. (1995). *Genius: The natural history of creativity.* Cambridge: Cambridge University Press.

Eysenck, H. J., & Barrett, P. (1985). Psychophysiology and the measurement of intelligence. In C. R. Reynolds & P. C. Wilson (Eds.), *Methodological and statistical advances in the study of individual differences* (pp. 1-49). New York: Plenum.

Fagan, B. M. (2001). *The little ice age: How climate made history, 1300– 1850.* New York: Basic Books.

Feldman, M., & Cavalli-Sforza L. L. (1976). Cultural and biological evolutionary processes, selection for a trait under complex transmission. *Theoretical Population Biology, 9,* 238-259.

Figuredo, A. J. (2012). Life history strategy and the evolution of eugenics. Invited Colloquium. Seminar on Chance Purpose and Progress in Evolution, Department of Ecology and Evolutionary Biology, University of Arizona, Tucson, Arizona.

Figueredo, A. J. (2009). Human capital, economic development, and evolution: A review and critical comparison of Lynn & Vanhanen (2006) and Clark (2007). *Human Ethology Bulletin, 24,* 5-8.

Figueredo, A. J., & Wolf, P. S. A. (2009). Assortative pairing and life history strategy: a cross-cultural study. *Human Nature, 20,* 317-330.

Flynn, J. R. (2009a). *What is intelligence? Beyond the Flynn effect* (expanded ed.). Cambridge: Cambridge University Press.

Flynn, J. R. (2009b). Requiem for nutrition as the cause of IQ gains: Raven's gains in Britain 1938–2008. *Economics and Human Biology, 7*, 18–27.

Flynn, J. R. (1987). Massive IQ gains in 14 nations: what IQ tests really measure. *Psychological Bulletin, 101*, 171–191.

Fogel, R. W. (1964). *Railroads and American economic growth: Essays in econometric history.* Baltimore: Johns Hopkins Press.

Fox, M. C., & Mitchum, A. L. (In press). A knowledge based theory of rising scores on "culture-free" tests. *Journal of Experimental Psychology: General.* doi:10.1037/a0030155.

Galton, F. (1869). *Hereditary genius.* London: MacMillan.

Galton, F. (1883). *Inquiries into human faculty and its development.* London: Dent.

Gary, B. L. (1993). *A new timescale for placing human events, derivation of per capita rate of innovation, and a speculation on the timing of the demise of humanity.* Unpublished Manuscript.

Gelade, G. A. (2008). IQ, cultural values, and the technological achievement of nations. *Intelligence, 36*, 711–718.

Giles-Bernadelli, B. M. (1950). The decline of intelligence in New Zealand. *Population Studies, 4*, 200-208.

Glad, J. (2006). *Future human evolution: Eugenics in the twenty-first century.* Shuylkill Haven: Hermitage Publishers.

Glad, J. (1998). A hypothetical model of IQ decline resulting from political murder and selective emigration in the former U.S.S.R. *Mankind Quarterly, 38*, 279–298.

Gottfredson, L. S. (1997). Mainstream science on intelligence: an editorial with 52 signatories, history, and bibliography. *Intelligence, 24*, 13-23.

Graham, R. K. (1998). Devolution by revolution: selective genocide ensuing from the French and Russian revolutions. *Mankind Quarterly, 39*, 71-93.

Hamilton, W. D. (2002). *Narrow roads to gene land, volume 2: Evolution of sex.* New York: Oxford University Press

Hamilton, W. D. (2000). A review of Dysgenics: Genetic Deterioration in Modern Populations. *Annals of Human Genetics, 64*, 363-374.

Hart, M. (2007). *Understanding human history: An analysis including the effects of geography and differential evolution.* Atlanta: Washington Summit Publishers.

Havender, W. R. (1987). Educational and social implications. In Modgil, S., & Modgil, C, eds. *Arthur Jensen, consensus and controversy* (pp. 397-411). Philadelphia: Routledge Falmer.

Haier, R. J., Siegel, B., Tang, C., Abel, L., & Buchsbaum, M. S. (1992). Intelligence and changes in regional cerebral glucose metabolic rate following learning. *Intelligence, 16*, 415-426.

Healey, D. (2013). *Pharmageddon.* Berkeley: University of California Press.

Herrnstein, R. J., & Murray, C. (1994). *The bell curve: Intelligence and class structure in American life.* New York: Free Press Paperbacks.

Higgins, J. V., Reed, E. W., & Reed, S. C. (1962). Intelligence and family size: a paradox resolved. *Eugenics Quarterly*, 9, 84-90.

Horgan, J. (1997). *The end of science: Facing the limits of knowledge in the twilight of the scientific age.* New York: Broadway Book.

Howard, R. W. (2001). Searching the real world for signs of rising population intelligence. *Personality and Individual Differences, 30*, 1039–1058.

Howard, R. W. (1999). Preliminary real-world evidence that average human intelligence really is rising. *Intelligence, 27*, 235–250.

Hu, L., & Bentler, P.M. (1999). Cutoff criteria for fit indexes in covariance structure analysis: conventional criteria versus new alternatives. *Structural Equation Modeling, 6*, 1-55.

Hu, L. T., & Bentler, P. (1995). Evaluating model fit. In Hoyle, R. H, ed. *Structural equation modeling. Concepts, issues, and applications.* (pp.76.99). London: Sage.

Huebner, J. (2005). A possible declining trend for worldwide innovation. *Technological Forecasting and Social Change, 72*, 980–986.

Hunter, J. E., & Schmidt, F. L. (2004). *Methods of meta-analysis (2nd Ed.): Correcting error and bias in research findings.* Thousand Oaks: Sage.

International Energy Agency. (2011). *CO_2 emissions from fuel combustion highlights.* Paris: OECD/IEA.

Itzkoff, S. W. (2009). *The end of economic growth.* New York: The Edwin Mellen Press.

James, L. R., Mulaik, S. A., & Brett, J. M. (1982). *Causal analysis: Assumptions, models, and data.* Beverly Hills: Sage.

Jensen, A. R. (2011). The theory of intelligence and its measurement. *Intelligence, 39*, 171-177.

Jensen, A. R. (2006). *Clocking the mind: Mental chronometry and individual differences.* Oxford: Elsevier.

Jensen, A. R. (1998). *The g factor: The science of mental ability.* Westport: Praeger.

Jensen, A. R. (1997). The puzzle of nongenetic variance. In: Sternberg, R. J., & Grigorenko, E. L, eds. *Heredity, intelligence, and environment.* (pp 42-88).Cambridge: Cambridge University Press.

Johnson, W., Bouchard Jr., T. J., McGue, M., Segal, N. L., Tellegen, A., Keyes, M., & Gottesman, I. I. (2007). Genetic and environmental influences on the Verbal-Perceptual-Image Rotation (VPR) model of the structure of mental abilities in the Minnesota study of twins reared apart. *Intelligence, 35*, 542-562

Jones, B. (2009). The burden of knowledge and the "death of the renaissance man": is innovation getting harder? *The Review of Economic Studies*, 76, 283–317.

Jones, B. F., & Weinberg, B. A. (2011). Age dynamics in scientific creativity. *Proceedings of the National Academy of Science, USA, 108,* 18910-18914.

Jordan, D. S. (1915). *War and the breed: The relation of war to the downfall of nations.* Boston: The Beacon Press.

Juan-Espinosa, M., Cuevas, L., Escorial, S., & García, L. F. (2006). The differentiation hypothesis and the Flynn effect. *Psicothema, 18,* 284–287.

Kane, H. D. (2000). A secular decline in Spearman's g: evidence from the WAIS, WAIS-R and WAIS-III. *Personality and Individual Differences, 29,* 561–566.

Kane, H. D., & Oakland, T. D. (2000). Secular declines in Spearman's g: some evidence from the United States. *The Journal of Genetic Psychology, 161,* 337–345.

Kealey, T. (1997). *The economic laws of scientific research.* London: Palgrave Macmillan.

Kasparov, G., Levchin, M., & Thiel, P. (2012). *The blueprint: Reviving innovation, rediscovering risk and rescuing the free market.* New York: W. W. Norton & Company.

Koriat, A. (2012). When are two heads better than one and why? *Science, 336,* 360-362.

Krischel, M. (2012). Perceived hereditary effects of World War I : a study of the positions of Freidrich von Bernhardi and Vernon Kellogg. *Medicine Studies, 2,* 139-150.

La Griffe du Lion. (2004). Smart fraction theory II: why Asians lag. 6(2): http://www.lagriffedulion.f2s.com/sft2.htm

Lentz, T. (1927). Relation of IQ to size of family. *Journal of Educational Psychology, 18,* 486–496.

Loehlin, J. C. (1997). Dysgenesis and IQ. *The American Psychologist*, 52, 1236–1239.

Lynn, R. (2012). IQs predict differences in technological development of nations from 1000 BC through 2000 AD. *Intelligence, 40*, 439-444.

Lynn, R. (2011). *Dysgenics: Genetic deterioration in modern populations* (Second revised edition). London: Ulster Institute for Social Research.

Lynn, R. (2009). What has caused the Flynn effect? Secular increases in the development quotients of infants. *Intelligence, 37*, 16–24.

Lynn, R. (2006). *Race differences in intelligence: An evolutionary perspective.* Atlanta: Washington Summit Press.

Lynn, R. (2001). *Eugenics: A Reassessment.* Westport: Praeger.

Lynn, R. (1999). New evidence for dysgenic fertility in intelligence in the United States. *Social Biology, 46*, 146–153.

Lynn, R. (1996). *Dysgenics: Genetic deterioration in modern populations.* Westport: Praeger.

Lynn, R. (1991). Race differences in intelligence: a global perspective. *Mankind Quarterly, 31*, 255-296.

Lynn, R. (1990). The role of nutrition in secular increases in intelligence. *Personality and Individual Differences, 11*, 273-285.

Lynn, R. (1989). A nutrition theory of the secular increases in intelligence — positive correlations between height, head size and IQ. *British Journal of Educational Psychology, 59*, 372–377.

Lynn, R. (1983). IQ in Japan and the United States shows a growing disparity. *Nature, 306*, 291–292.

Lynn, R., & Cooper, C. (1993). A secular decline in Spearman's *g* in France. *Learning and Individual Differences, 5*, 43–48.

Lynn, R., & Cooper, C. (1994). A secular decline in the strength of Spearman's *g* in Japan. *Current Psychology, 13*, 3–9.

Lynn, R., & Harvey, J. (2008). The decline of the world's IQ. *Intelligence, 36*, 112–120.

Lynn, R., & van Court, M. (2004). New evidence of dysgenic fertility for intelligence in the United States. *Intelligence, 32*, 193–201.

Lynn, R., & Vanhanen, T. (2012). *Intelligence: A unifying construct for the social sciences.* London: Ulster Institute for Social Research.

Mackintosh, N. J (2002). Dysgenics: Genetic Deterioration in Modern Populations. By Richard Lynn. pp. 237. (Praeger, 1996.) £48.95, 0-275-94917-6, hardback. *Journal of Biosocial Science, 34*, 283–284.

Maddox, J. (1999). *What remains to be discovered: Mapping the secrets of the universe, the origins of life, and the future of the human race.* New York: Free Press Touchstone.

Maxwell, J. (1954). Intelligence, fertility and the future: a report on the 1947 Scottish Mental Survey. *Eugenics Quarterly, 1*, 244–247.

McCurdy, H. G. (1960). The childhood pattern of genius. *Horizon, 2*, 33-38.

Meisenberg, G. (2010). The reproduction of intelligence. *Intelligence, 38*, 220–230.

Meisenberg, G. (2009). Wealth, intelligence, politics and global fertility differentials. *Journal of Biosocial Sciences, 41*, 519–535.

Meisenberg, G. (2008). How universal is the negative correlation between education and fertility? *Journal of Social, Political and Economic Studies, 33*, 205–227.

Meisenberg, G. (2007). *In God's image. The natural history of intelligence and ethics.* Brighton: Guild Publishing.

Meisenberg, G. (2003). IQ population genetics: it's not as simple as you think. *Mankind Quarterly, 44*, 185-210.

Meisenberg, G., & Kaul, A. (2010). Effects of sex, race, ethnicity and marital status on the relationship between intelligence and fertility. *Mankind Quarterly, 50*, 151–187.

Meisenberg, G., Lawless, E., Lambert, E., & Newton, A. (2006). The social ecology of intelligence on a Caribbean island. *Mankind Quarterly, 46*, 395-433.

Meisenberg, G., Lawless, E., Lambert, E., & Newton, A. (2005). The Flynn effect in the Caribbean: generational change of cognitive test performance in Dominica. *Mankind Quarterly, 46*, 29–69.

Miller, G. F., & Penke, L. (2007). The evolution of human intelligence and the coefficient of additive genetic variance in human brain size. *Intelligence, 35*, 97-114.

Mingroni, M. A. (2007). Resolving the IQ paradox: heterosis as a cause of the Flynn Effect and other trends. *Psychological Review, 114*, 806–829.

Mingroni, M. A. (2004). The secular rise in IQ: giving heterosis a closer look. *Intelligence, 32*, 65–83

Modis, T. (2005). Comments on the Huebner article. *Technological Forecasting and Social Change, 72*, 987–988.

Moshinsky, P. (1939). The correlation between fertility and intelligence within social classes. *Sociological Review, 31*, 144-165.

Murray, C. (2012). *Coming apart: The state of white America, 1960-2010*. New York: Crown Publishing Group.

Murray, C. (2003). *Human accomplishment: The pursuit of excellence in the arts and sciences, 800 BC to 1950*. New York: Harper Collins.

Must, O., Must, A., & Raudik, V. (2003a). The secular rise in IQs: in Estonia the Flynn effect is not a Jensen effect. *Intelligence, 167*, 1–11.

Must, O., Must, A., & Raudik, V. (2003b). The Flynn Effect for gains in literacy found in Estonia is not a Jensen Effect. *Personality and Individual Differences, 34*, 1287-1292.

Must, O., te Nijenhuis, J., Must, A., & van Vianen, A. E. M. (2009). Comparability of IQ scores over time. *Intelligence, 37*, 25-33.

Nagoshi, G. T., & Johnson, R. C. (1986). The ubiquity of g. *Personality and Individual Differences, 7,* 201-207.

Neisser, U. (ed,). (1997). *The rising curve: Long-term gains in IQ and related measures.* Washington DC: American Psychological Association.

Neisser, U., Boodoo, G., Bouchard, T. J. Jr., Boykin, A. W., Brody, N., Ceci, S. J., Halpern, D. F., Loehlin, J. C., Perloff, R., Sternberg, R. J., & Urbina, S. (1996). Intelligence: knowns and unknowns. *American Psychologist, 51,* 77–101.

Nettelbeck, T., & Wilson, C. (2004). The Flynn effect: smarter not faster. *Intelligence, 32,* 85-93.

Nevin, R. (2000). How lead exposure relates to temporal changes in IQ, violent crime, and unwed pregnancy. *Environmental Research, 83,* 1-22.

Niosi, J., ed. (1991). *Technology and national competitiveness: Oligopology, technological innovation and international competitiveness.* Montreal: McGill-Queen's University Press.

Nisbet, J. D. (1958). Intelligence and family size, 1949-56. *Eugenics Review, 49,* 4-5.

Nowak, M. A., Tarnita, C. E., & Wilson, E. O. (2010). The evolution of eusociality. *Nature, 466,* 1057-1062.

Nyborg, H. (2012). The decay of Western civilization: double relaxation of Darwinian selection. *Personality and Individual Differences, 53,* 118-125.

Nyborg, H. (2005). Sex related differences in general intelligence g, brain size and social status. *Personality and Individual Differences, 39,* 497–509.

Oesterdiekhoff, G. W. (2012). Was pre-modern man a child? The quintessence of the psychometric and developmental approaches. *Intelligence, 40,* 470-478.

Panagopoulos, A. (2004). When does patent protection stimulate innovation? Discussion Paper No. 04/565.

Papavassiliou, I. T. (1954). Intelligence and family size. *Population Studies, 7*, 222-226.

Paxton, C. G. M. (1998). A cumulative species description curve for large open water marine animals. *Journal of the Marine Biological Association of the UK, 78*, 1389-1391.

Pearson, K. (1903). On the inheritance of the mental and moral characters in man. *Journal of the Anthropological Institute of Great Britain and Ireland, 33*, 179-237.

Pinker, S. (2011). *The better angels of our nature: Why violence has declined.* New York: Viking Adult.

Plomin, R., De Fries, J. C., & McClearn, G. E. (1990). *Behavioral genetics.* New York: Freeman.

Pugh, M. (1999). *Britain since 1789: A Concise History.* New York: Macmillan.

Prokosch, M. D., Yeo, R. A., & Miller, G. F. (2005). Intelligence tests with higher *g*-loadings show higher correlations with body symmetry: evidence for a general fitness factor mediated by developmental stability. *Intelligence, 33*, 203-213.

Qian, Y. (2007). Do national patent laws stimulate domestic innovation in a global patenting environment? A cross-country analysis of pharmaceutical patent protection, 1978-2002. *Review of Economics and Statistics, 89*, 436-453.

Rae, C., Scott, R. B., Thompson, C. H., Kemp, G. J., Dumughn, I., Styles, P., Tracey, I., & Radda, G. K. (1996). Is pH a biochemical marker of IQ? *Proceedings of the Royal Society Series B: Biological Sciences, 263*, 1061-1064.

Reeve, C. L. (2009). Expanding the *g*-nexus: further evidence regarding the relations among national IQ religiosity and national health outcomes. *Intelligence, 37*, 495–505.

Retherford, R. D., & Sewell, W. H. (1988). Intelligence and family size reconsidered. *Social Biology, 35*, 1–40.

Ricardo, D. (1891). The principles of political economy. In: Gonner, E. C. K ed. *Bohn's Economic Library*. London: Bell & Sons.

Richerson, P. J., & Boyd, R. (2006). *Not by genes alone: How culture transformed human evolution*. Chicago: University of Chicago Press.

Rijsdijk, F. V., Vernon, P. A., & Boomsma, D. I. (1998). The genetic basis of the relation between speed-of-information-processing and IQ. *Behavioral and Brain Research, 95*, 77-84.

Rindermann, H., Sailer, M., & Thompson, J. (2009). The impact of smart fractions, cognitive ability of politicians and average competence of peoples on social development. *Talent Development and Excellence, 1*, 3-25.

Rindermann, H., & Thompson, J. (2011a). Intelligence of the future and economic development. Paper presented at the 12th Annual Meeting of the International Society for Intelligence Research, Lanarca, Cyprus.

Rindermann, H., & Thompson, J. (2011b). Cognitive capitalism: the effect of cognitive ability on wealth, as mediated through scientific achievement and economic freedom. *Psychological Science, 6*, 754-763.

Roberts, J. A. F., Norman, R. M., & Griffiths, R. (1938). Studies on a child population: Intelligence and family size. *Annals of Eugenics, 8*, 178-215.

Robertson, K. F., Smeets, S., Lubinski, D., & Benbow, C. P. (2010). Beyond the threshold hypothesis: even among the gifted and top math/science graduate students, congnitive abilities, vocational interests, and lifestyle preferences matter for career choice, performance, and persistence. *Current Directions in Psychological Science, 19*, 346-351.

Rowe, D. C., & Rodgers, J. L. (2002). Expanding variance and the case of historical changes in IQ means: a critique of Dickens and Flynn (2001). *Psychological Review, 109*, 759–763.

Rushton, J. P. (1999). Secular gains not related to the *g* factor and inbreeding depression – unlike Black – White differences: a reply to Flynn. *Personality and Individual Differences, 26*, 381-389.

Rushton, J. P. (1998). The "Jensen Effect" and the "Spearman-Jensen hypothesis" of Black-White IQ differences. *Intelligence, 26*, 217-225.

Rushton, J. P., & Ankney, C. D. (2009). Whole brain size and general mental ability: a review. *International Journal of Neuroscience, 119*, 692-732.

Rushton, J. P., & Jensen, A. R. (2010). The rise and fall of the Flynn effect as a reason to expect a narrowing of the Black-White IQ gap. *Intelligence, 38*, 213-219.

Saleeby, C. W. (1914). *The progress of eugenics*. London: Methuen.

Saleeby, C. W. (1913). Eugenics and dysgenics in relation to alcohol. *The British Journal of Inebriety, 11*, 1-7.

SAS Institute Inc. (2011). *Base SAS ® 9.3 procedures guide: Statistical procedures*. Cary: SAS Institute Inc.

Schafer, E. W. P. (1985). Neural adaptability: a biological determinant of *g* factor intelligence. *Behavioral and Brain Sciences, 8*, 240-241.

Science Service (Crew, F. A. E., Darlington, C. D., Haldane, J. B. S., Harland, S. C., Hogben, L. T., Huxley, J. S., Muller, H. J., Needham, J., Chid, G. P., David, P. R., Dahlberg, G., Dobzhansky, T., Emerson, R. A., Gordon, C., Hammond, J., Huskins, C. L., Koller, P. C., Landauer, W., Plough, H. H., Price, B., Schultz, J., Steinberg, A. G., & Waddington, C. H.). (1939). Social biology and population improvement. *Nature, 411*, 521-522.

Shatz, S. M. (2008). IQ and fertility: a cross-national study. *Intelligence, 36*, 109-111.

Shayer, M., & Ginsburg, D. (2009). Thirty years on — A large anti-Flynn effect? (II): 13- and 14-year olds. Piagetian tests of formal operations norms 1976–2006/7. *British Journal of Educational Psychology, 79*, 409–418.

Shayer, M., Ginsburg, D., & Coe, R. (2007). Thirty years on – a large anti-Flynn effect? The Piagetian test of Volume & Heaviness norms 1975-2003. *British Journal of Educational Psychology, 77*, 25-41.

Silverman, I. W. (2010). Simple reaction time: it is not what it used to be. *American Journal of Psychology, 123*, 39–50.

Simonton, D. K. (2013). After Einstein: scientific genius is extinct. *Nature, 493*, 602.

Simonton, D. K. (2009). Varieties of (scientific) creativity: A hierarchical model of domain-specific disposition, development, and achievement. *Perspectives on Psychological Science, 4*, 441–452.

Simonton, D. K. (2003). Exceptional creativity across the life span: The emergence and manifestation of creative genius. In L. V. Shavinina (Ed.), *The International Handbook of Innovation* (pp. 293-308). New York: Pergamon Press.

Simonton, D. K. (1999). *Origins of genius: Darwinian perspectives on creativity.* New York: Oxford University Press.

Skirbekk, V. (2008). Fertility trends by social status. *Demographic Research, 18*, 145-180.

Smith, S. (1942). Language and nonverbal test performance of racial groups in Honolulu before and after a 14 year interval. *Journal of General Psychology, 26*, 51-93.

Spearman, C. (1904). "General intelligence," objectively determined and measured. *American Journal of Psychology, 15*, 201–293.

Spearman, C. (1927). *Abilities of man: Their nature and measurement.* New York: Macmillan.

Sternberg, R. J. (2005). There are no public policy implications: a reply to Rushton and Jensen (2005). *Psychology, Public Policy and Law, 11,* 295-301.

Sundet, J. M., Barlaug, D. G., & Torjussen, T. M. (2004). The end of the Flynn effect? A study of secular trends in mean intelligence test scores of Norwegian conscripts during half a century. *Intelligence, 32,* 349–362.

Sundet, J. M., Tambs, K., Magnus, P., & Berg, K. (2002). On the question of secular trends in the heritability of intelligence test scores: a study of Norwegian twins. *Intelligence, 12,* 47-59.

Sutherland, H. E. G. (1930). The relationship between intelligence quotient and size of family in the case of fatherless children. *Journal of Genetic Psychology, 38,* 161-170.

Sutherland, H. E. G. & Thomson, G. H. (1926). The correlation between intelligence and size of family. *British Journal of Psychology, 17,* 81-92.

Teasdale, T. W., & Owen, D. R. (2008). Secular declines in cognitive test scores: a reversal of the Flynn Effect. *Intelligence, 36,* 121-126.

Teasdale, T. W., & Owen, D. R. (2005). A long-term rise and recent decline in intelligence test performance: the Flynn Effect in reverse. *Personality and Individual Differences, 39,* 837-843.

te Nijenhuis, J. (In press). The Flynn effect, group differences and *g* loadings. *Personality and Individual Differences.* doi:10.1016/j.paid.2011.12.023

te Nijenhuis, J., & van der Flier, H. (2007). The secular rise in IQs in the Netherlands: is the Flynn effect on g? *Personality and Individual Differences, 43,* 1259-1265.

te Nijenhuis, J., & van der Flier, H. (In press). Is the Flynn effect on *g*?: a meta-analysis. *Intelligence.* Doi: 10.1016 /j.intell.2013.03.001.

te Nijenhuis, J., van Vianen, A. E. M., & van der Flier, H. (2007). Score gains on *g*-loaded tests: no g. *Intelligence, 35,* 283-300.

Thomson, G. H. (1949). Intelligence and fertility: the Scottish 1947 survey. *Eugenics Review, 41,* 163-170.

Thomson, G. H. (1946). The trend of national intelligence. *Eugenics Review, 38,* 9-18.

Thurstone, L. L., & Jenkins, R. L. (1931). *Order of birth, parental age and intelligence.* Chicago: University of Chicago Press.

Trzaskowski, M., Yang, J., Visscher, P. M., & Plomin, R. (In press). DNA evidence for strong genetic stability and increasing heritability of intelligence from age 7 to 12. *Molecular Psychiatry.* doi:10.1038/mp.2012.191.

Tuddenham, R. D. (1948). Soldier intelligence in World Wars I and II. *The American Psychologist, 3,* 54–56.

Upton, G., & Cook, I. (2006). *Oxford dictionary of statistics.* Oxford: Oxford University Press.

van Bloois, R. M., Geutjes, L. L., te Nijenhuis, J., & de Pater, I. E. (2009). *g* loadings and their true score correlations with heritability coefficients, giftedness, and mental retardation: three psychometric meta-analyses. Paper presented at the 10th Annual Meeting of the International Society for Intelligence Research, Madrid, Spain, December.

van Court, M., & Bean, F. D. (1985). Intelligence and fertility in the United States: 1912–1982. *Intelligence, 9,* 23–32.

Vernon, P. E. (1951). Recent investigations of intelligence and its measurement. *Eugenics Review, 43,* 125-137.

Vining, D. R. (1982). On the possibility of the re-emergence of a dysgenic trend with respect to intelligence in American fertility differentials. *Intelligence, 6,* 241–264.

Vining, D. R. (1995). On the possibility of the reemergence of a dysgenic trend with respect to intelligence in American fertility differentials: an update. *Personality and Individual Differences, 19*, 259–263.

Weiss, V. (2007). The population cycle drives human history from a eugenic phase into a dysgenic phase and eventual collapse. *Journal of Social Political and Economic Studies, 32*, 327–358.

Wicherts, J. M., Dolan, C. V., Hessen, D. J., Oosterveld, P., van Baal, G. C. M., Boomsma, D. I., & Span, M. M. (2004). Are intelligence tests measurement invariant over time? Investigating the nature of the Flynn effect. *Intelligence, 32*, 509–537.

Wilson, D. S. (2002). *Darwin's cathedral: Evolution, religion, and the nature of society.* Chicago: University of Chicago Press.

Wolf, P. S. A., & Figueredo, A. J. (2011). Fecundity, offspring longevity, and assortative mating: Parametric tradeoffs in sexual and life history strategy. *Biodemography and Social Biology, 52*, 171-183.

Woodley, M. A. (2012a). The social and scientific temporal correlates of genotypic intelligence and the Flynn effect. *Intelligence, 40*, 189-204.

Woodley, M. A. (2012b). A life history model of the Lynn-Flynn effect. *Personality and Individual Differences, 53*, 152-156.

Woodley, M. A. (2011a). The cognitive differentiation-integration effort hypothesis: a synthesis between the fitness indicator and life history models of human intelligence. *Review of General Psychology, 15*, 228-245.

Woodley, M. A. (2011b). Heterosis doesn't cause the Flynn effect: a critical examination of Mingroni (2007). *Psychological Review, 118*, 689-693.

Woodley, M. (2006). The role of niche construction in the evolution and future of the human brain. *Mankind Quarterly, 46*, 461-477.

Woodley, M. A., Figueredo, A. J., Brown, S. D. & Ross, K. C (In press). Four successful tests of the cognitive differentiation-integration effort hypothesis. *Intelligence.* Doi:10.1016/j.intell.2013.02.002

Woodley, M. A., & Madison, G. (In press). Establishing an association between the Flynn effect and ability differentiation. *Personality and Individual Differences.* Doi:10.1016/j.paid.2013.03.016.

Woodley, M. A., & Meisenberg, G. (In press). A Jensen effect on dysgenic fertility: an analysis involving the National Longitudinal Survey of Youth. *Personality and Individual Differences.* doi:10.1016/j.paid.2012.05.024

Woodley, M. A., & Meisenberg, G. (2013). In the Netherlands the anti-Flynn effect is a Jensen effect. *Personality and Individual Differences, 54*, 871-876.

Zhang, D. D., Brecke, P., Lee, H. F., He, Y-Q., & Zhang, J. (2007). Global climate change, war, and population decline in recent human history. *Proceedings of the National Academy of Sciences, 104*, 19214-19219.

Zhang, D. D., Lee, H. F., Wang, C., Li, B., Pei, Q., Zhang, J., & An, Y. (2011). The causality analysis of climate change and large-scale human crises. *Proceedings of the National Academy of Sciences, 108*, 17296-17301.

www.ingramcontent.com/pod-product-compliance
Lightning Source LLC
Chambersburg PA
CBHW021823190326
41518CB00007B/724